Light Infantry of the Greek and Roman World

Light Infantry of the Greek and Roman World

Forgotten Heroes of Ancient Warfare

Jean Charl Du Plessis

Pen & Sword
MILITARY

First published in Great Britain in 2024 by
Pen & Sword Military
An imprint of Pen & Sword Books Limited
Yorkshire – Philadelphia

ISBN 978 1 39908 148 1

A CIP catalogue record for this book is available from the British Library

Typeset by Mac Style
Printed in the UK by CPI Group (UK) Ltd, Croydon, CR0 4YY.

MIX
Paper | Supporting responsible forestry
FSC® C013604

Pen & Sword Books Limited incorporates the imprints of After the Battle, Atlas, Archaeology, Aviation, Discovery, Family History, Fiction, History, Maritime, Military, Military Classics, Politics, Select, Transport, True Crime, Air World, Frontline Publishing, Leo Cooper, Remember When, Seaforth Publishing, The Praetorian Press, Wharncliffe Local History, Wharncliffe Transport, Wharncliffe True Crime and White Owl.

For a complete list of Pen & Sword titles please contact

PEN & SWORD BOOKS LIMITED
47 Church Street, Barnsley, South Yorkshire, S70 2AS, England
E-mail: enquiries@pen-and-sword.co.uk
Website: www.pen-and-sword.co.uk
or
PEN AND SWORD BOOKS
1950 Lawrence Rd, Havertown, PA 19083, USA
E-mail: uspen-and-sword@casematepublishers.com
Website: www.penandswordbooks.com

Contents

Introduction

Tales of ancient battlefields conjure up images of ferocious and bloody battles: serried ranks of phalanxes and maniples of thousands of metal-clad warriors clashing on flat, open plains. Warriors thrusting with ash spears and monstrous pikes and wielding deadly swords from behind a wall of heavy shields. These battles were intense, fierce and violent. Soldiers forcefully pushed against one another, trampling fields covered in blood and gore. The screams of the dying and wounded mingled with the crash of metal. On the flanks, horses were spurred to a gallop as their riders slashed from high above their heads with long, glistening slashing swords or charged with couched lances. Large-scale pitched battles of the ancient world were a terrible thing to behold, but they were few and far between.

These pitched battles, where thousands of men simultaneously collided and slaughtered one another for dominance of the field, played only a small part in ancient warfare. While the outcome of wars was often based on these climactic large-scale clashes between heavy armoured infantry and cavalry, these battles were culminations of an often drawn-out, taxing and carefully planned campaign. The heavy burden of the success of these campaigns did not, however, fall on the heavy troops, but on the more lightly armed auxiliary forces. These light infantry were often disregarded by ancient authors in their accounts of warfare, since they lacked the illustrious reputation of the heavy infantry and cavalry. An example of this is Thucydides' description of the campaign of Athenian general Demosthenes in Aetolia. Thucydides is known for his meticulously detailed and accurate account of the Peloponnesian War, and during particular campaigns he gives the number of heavy infantry and provides the names of their officers but fails to mention even the number of light troops (on this particular occasion, archers), referring to the captain of the 'lights' simply as 'the captain'.[1] It must be noted that the service of these archers in the campaign was exemplary, serving as the only defence for Demosthenes' heavy infantry against the enemy's javelins until their arrows ran out. Once their arrows were spent, the enemy light troops devastated the Athenian hoplites and completely routed them, killing 'by far the best men in the city of Athens'.[2]

Light infantry were, more often than not, drawn from the poorest members of society and subjected vassals, recruited from the fringes of society and 'barbarian' peoples, or they came from the scorned profession of mercenaries. Furthermore, they did not stand their ground in the face of the enemy onslaught, instead often fighting from a distance. They did not look their enemy in the eye before delivering their mortal blow, but did their killing from afar. These were hardly the heroic deeds of valour, manliness and daring sought after by the historian's pen. Modern scholarship has followed in this trend, and very few publications in more recent literature are devoted to light-armed troops. Although their role has often been overlooked, they played an intricate and essential role in ancient warfare.

War campaigns in the ancient world consisted of long, burdensome marches where the protection and defence, not only of the baggage train and supply lines but of the heavy troops themselves, relied on the light troops. For these marches to be conducted safely, scouting was required to avoid ambushes and find safe passages through mountains and other difficult terrain. Enemy positions and battlefields also had to be scouted beforehand. These scouting duties fell on the light troops too. There were intense forced marches, where the equipment of the heavy infantry slowed them down too much for the speedy manoeuvres required over great distances, so it was the light troops who embraced this role. Light troops were also particularly suited to urban warfare, where crooked and narrow streets severely restricted the formations of the heavy infantry.

Pitched battles on the open plain were a rarity, but ancient campaigns witnessed countless smaller, non-pitched battles and skirmishes, often in rugged and broken terrain where the heavy troops could not operate, and this fighting had to be done by the light troops. The sieges of cities and strongholds saw much of the action conducted by the light troops, and any battle scenario that required ranged warfare fell to them too. The capture of strategic positions such as watering holes, river crossings and high ground required speed, flexibility and manoeuvrability, meaning such duties were also designated to lightly armed soldiers. Guerrilla warfare, ambuscades, any type of hit-and-run tactics and the harassment of enemy formations, as well as raiding and plundering of enemy territory, were further scenarios ideally performed by light troops. In all these cases, the speed and mobility needed for these operations fell outside of the ability of the inflexible heavy troops.

Even though the role of light troops in pitched battles was overshadowed by that of their more heavily armed compatriots, they still played important parts in these large-scale encounters. Indeed, the opening stages of these battles were initiated by the light troops, who manned a skirmishing screen to defend the heavy infantry as they assembled in their formations to prepare for the main clash. During these early stages of the battle, they generally skirmished against the enemy's light troops while also seeking to disrupt the enemy's main formation and weaken their morale. When the actual battle started, the light troops often took position on the flanks of the heavy troops, acting as a second shield for the massed infantry and protecting their vulnerable flanks. The lightly armed soldiers also operated in a more offensive capacity by acting as support for the cavalry attacking the enemy flanks, pursuing fleeing opponents to ensure they were completely routed from the field and capturing many prisoners.

Because of the nature of their style of warfare, mobility was the light troops' greatest advantage. They wore little or no armour to remain as agile as possible, and their weapons also reflected their style of warfare. Because of their lack of armour, the light troops generally avoided hand-to-hand combat and melee engagements. Their offensive weapons were largely range weapons: the bow, javelin and sling. Their defensive weapons, if they had any, were bucklers and small light shields which would not hamper their mobility. The lower classes, because of their financial restriction and the expense of heavy armour, were naturally allocated to these combat roles. Light troops were often hired as mercenaries from peoples whose native land made them natural light infantry, being mountainous for example. Light troops were also obtained from peoples where a certain type of weapon or fighting style was endemic to their culture and way of life. This often led to certain types of light troops filling specialized roles and made them highly sought after throughout the Mediterranean world and beyond. The feared Cretans archers and deadly Balearic slingers serve as prime examples.

While the heavy troops were the 'glory boys' of the ancient battlefield, whose names and deeds are forever inked into the pages of history (battles such as Thermopylae, Gaugamela, Cannae and Zama come to mind), the success of these soldiers was built on the labours and blood of the light troops. They were more than just a means to swell the ranks of ancient armies, and should instead be viewed as an effective answer to the limitations of the heavy troops, forming an intricate and indispensable part of the army. The great Athenian general Iphicrates made the interdependency of a combined army crystal clear:

> 'The light-armed troops are like the hands, the cavalry like the feet, the line of men-at-arms itself like chest and breastplate.'[3]

This book explores the light infantry of the ancient Graeco-Roman world, from Archaic Greece to the end of the Roman Republic, seeking to bring to the forefront these unsung heroes of ancient times and highlight the essential role they played in Greek and Roman warfare. It examines light troop terminology used by the Greek and Roman military writers, the various troop types which were considered 'light infantry', their various arms, armour and tactics, and how the light infantry operated on campaign and in battles.

Chapter 1

Identifying the Light Infantry

1.1 The Term 'Light' as a Tactical Function

Accounts of Graeco-Roman warfare have always clearly distinguished between 'light' and 'heavy' infantry. The heavy infantry relied on disciplined ranks, fighting shoulder-to-shoulder from behind heavy shields, often clad in heavy armour. Their role was to meet the enemy heavy forces head-on in close combat during pitched battles. Their arms and armour provided them with the staying power for prolonged melee engagements. In addition to his protective gear, the heavy infantryman relied on the depth of the tactical formation and on his companions' fighting prowess, discipline, training and unwavering commitment, to keep his place in the formation. Cohesion of the heavy infantry formation was its greatest strength, being vital to their success. The term 'heavy' is attributed to these soldiers based on their function just as much as on their armament; perhaps even more so.

The light infantry stood in stark contrast to the heavy infantry. They fought as individuals, as opposed to the well-formed ranks of the heavy troops. They wore little if any armour, relying on the speed and agility this afforded them to keep safe. The lack of heavy armour was the core element which dictated their tactics and function. Because of the minimal defensive equipment, light troops were vulnerable in hand-to-hand combat and therefore favoured fighting from a distance. The bow, javelin and sling were thus their weapons of choice and they showered the enemy with various kinds of missiles, withdrawing before the enemy could engage them. Furthermore, the agility and speed that the lack of heavy equipment afforded them allowed the light troops to operate in rough and difficult terrain. Since they did not fight in set formations, rough terrain did not break up their ranks. Their light equipment allowed them the nimbleness to easily navigate around and over obstacles that uneven terrain could pose. On the battlefield, their light gear also lent certain tactical advantages, providing them with the swiftness and manoeuvrability to outflank enemy formations and hide themselves in ambuscades. Although lightly armed soldiers generally preferred to fight from a distance, these tactical advantages did allow them to engage in melees at times, such as when charging the enemy flank or rear.

The term 'light' attributed to these troops, as with the heavy troops, was based on their function, tactical mobility and deployment, as well as their equipment.

1.2 Light Infantry Terminology – Collective Terms

For our information on light infantry during the period, we rely predominantly on Greek and Roman historiography, and there is a variety of terms used by the ancient sources to describe light infantry. Military terminology among the historiographers of the period was by no means standard. Some words were used as collective or umbrella terms, such as *psiloi*, *euzōnoi* and *levis armatura*, which cover all manner of light troops, while other terms were more specific and defined. Terms such as *peltastai* meant something very different in the fifth and fourth century BC compared to the third and second century BC. Ancient authors often used military terms in quite a liberal manner. Livy, for instance, uses the same word, *caetrati*, for *sarissa* pike-wielding heavy infantry of the Macedonian phalanx as he does for light troops in the Seleucid and Achaean armies who were armed in the Classical Greek *peltast* manner.[1] Furthermore, Livy and Caesar use this same term for the Spanish Iberian light infantry armed with small round buckler shields.[2]

The following section explores Greek and Roman light infantry terminology used in the ancient Graeco-Roman sources. It is a brief assessment of the various collective terms which refer to light troops in general, including how these terms were used by the ancient writers and how they may be interpreted by modern readers.

1.2.1 Psiloi

The term *psiloi* (ψιλοί), meaning 'bare', is often used by the Greek sources as a general term to describe unarmoured light troops who fought with bows, slings and javelins. It is, for instance, the term used by Asclepiodotus, Aelian and Arrian in their tactical military manuals as a collective term for light troops armed with range weapons.[3] Asclepiodotus describes the *psiloi* as follows:

> 'The corps of light infantry (*psiloi*) on the contrary [to the heavy infantry] uses the lightest equipment because it shoots from a distance, and is provided with neither greaves nor cuirasses, but with javelins and slings, and in general with these missiles which we call "long-distance missiles".'[4]

Dionysius of Halicarnassus also uses *psiloi* for light infantry (in this case, Roman troops at the Battle of Asculum) armed with range weapons:

> 'Furthermore, standing on the wagons, which were four-wheeled, were many also of the light-armed troops (*psiloi*) – bowmen, hurlers of stones and slingers.'[5]

The term was first used in a military context to describe light-armed troops during the fifth century BC by Herodotus in his *Persian Wars*.[6] The word could, however, at times refer to train attendants, waggoners, orderlies and other non-combatants, and thus served as an umbrella term for unarmoured men.[7] In a military context, the literature seems to indicate that the *psiloi* were unarmoured (or at most very lightly armoured) light infantry who fought only with range weapons. The troop types who fell under this term would be bowmen, javelinists and slingers, but also those who simply threw rocks or stones by hand. The word is used to describe light troops from various cultures throughout the Graeco-Roman period by a variety of authors who wrote in Greek. The term *psiloi* to denote light infantry appears in the texts of Greek historiographers such as Thucydides and Xenophon, the military manuals of Onasander, Arrian, Appian and Asclepiodotus, the biographies of Plutarch, the world history of Diodorus Siculus and Polybius' Roman-focused world history.[8] The term survived well into the Middle Ages, where it still referred to light infantry armed with range weapons in the Eastern Roman/Byzantine armies.[9]

Thucydides also refers to the *psiloi* as range infantry (archers, javelineers and slingers) and paints a vivid picture of their effectiveness and tactics:

> '[*Psiloi*] occupy the highest points in order to paralyze the enemy by surrounding him on every side and thus leaving him without any tangible adversary, exposed to the cross-fire of their host; plied by those in his rear if he attacked in front, and by those on one flank if he moved against those on the other. In short, wherever he went he would have the assailants behind him, and these light-armed assailants (*psiloi*), the most awkward of all; arrows, darts, stones, and slings making them formidable at a distance, and there being no means of getting at them at close quarters, as they could conquer flying, and the moment their pursuer turned they were upon him.'[10]

In the above quotation the advantages of light troops are clearly demonstrated. Due to their light equipment and not having to depend on disciplined formations, *psiloi* were able to take advantage of the terrain, 'occupy the highest points' and outmanoeuvre the enemy, 'surrounding him on every side'. The speed and agility afforded them and the fact that they fought from afar made them 'formidable at a distance' and difficult, if not impossible, for heavy infantry to catch and engage, 'thus leaving him [the heavy infantry] without any tangible adversary'.

Onasander, the first-century Greek philosopher who wrote the *Strategikon* (*Στρατηγικός*), is one of the most important sources on ancient military matters, providing us with information not always available in other sources. This is particularly true for light infantry tactics. The term *psiloi* appears several times within the *Strategikon*, and as with the tactical manuals, it is used to refer to archers, slingers and javelin men:

> 'The general will assign his light-armed troops (*psiloi*) – javelin-throwers, bowmen, and slingers – to a position in front of the phalanx.'[11]

Onasander highlights the main purpose of light infantry as to skirmish before the main battlelines meet in pitched battle, then at the start of the battle to take position on the wings so that they may attack the enemy's vulnerable flanks:

> 'If each army should have a number of light-armed troops, the general should order his own light-armed men to be the first to hurl their weapons against their opponents before the hand-to-hand battle; or after the clash of the phalanx, attacking from the flank, they should make use of their missiles, for thus the enemy will be forced together into a narrow space and will be greatly confused by such tactics.'[12]

Onasander, in his *Strategikon*, also mentions the role of the *psiloi* to forage and emphasizes their effectiveness on uneven broken terrain:

> 'If the battle should happen to be in a country that is level in some places but hilly in others, then the light-armed troops should by all means be stationed in the uneven section, and then, if the general himself should possess the heights [and] the plain and some part of the enemy's phalanx should possess the heights, he should send against them the light-armed troops; for from the uneven ground they can more easily hurl their weapons and retreat, or they can very easily charge up the slopes, if they are agile.'[13]

It is interesting to note that Polybius, who is one of the most important sources on military matters of the Hellenistic period – particularly Roman history – rarely uses the term *psiloi* to describe light troops. On the few occasions where he does use the word, it refers to light infantry in Aetolian (4.12.4), Achaean (4.12.6), Macedonian, (5.7.11), Carthaginian (3.104.4) and Roman (3.104.6) armies. It should be noted that even when *psiloi* is used by Polybius to describe light troops, these very same troops are often labelled as *euzōnoi* later in the passage.[14] Polybius used the term *psiloi* more to describe open or 'bare' ground, such as the plain of Cannae, or the clear, open, 'bare' ground needed for the phalanx to operate on.[15] Polybius far preferred the term *euzōnoi* to describe the light troops from the Graeco-Roman world.

1.2.2 Euzōnoi

According to Daly, Polybius preferred the term *euzōnoi* as it was a more encompassing word which referred to the *psiloi*, the light troops who only fought with ranged weapons, and also light troops who could engage in hand-to-hand combat when the right opportunity presented itself, such as peltasts, the Spanish *caetrati*, *thōrakitai*, *thureophroi* and the Roman *velites*.[16] Scott and Liddell, in their Greek–English lexicon, define the term *euzōnoi* (εὔζωνοι) as 'active', 'unencumbered', 'well-girdled' or 'well-equipped'.[17] The term *euzōnoi* dates back to Homeric times as an epithet to describe 'well-girdled' women.[18] While the word appears in Herodotus and Thucydides, it is never in a military context. It is used to indicate the distance from one place to another by saying the *euzōnoi* ('well-equipped' or 'active' men) can make a journey in so many days.[19] From the fourth century BC, however, the term starts to appear in a military context in the writings of Xenophon to denote soldiers.[20] The word *euzōnoi* is used by Xenophon to describe infantry who detached from their divisions to force-march and rapidly capture strategic positions, who lent support to cavalry or who fought from afar with missiles.[21] On one occasion, Xenophon specified that these *euzōnoi* must be in the prime of their youth:

> 'Then Xenophon gave orders that the active men (*euzōnoi*) up to thirty years of age should move up from their several companies to the front. So he himself ran along with them, while Cleanor led the rest.'[22]

It is interesting to note that Xenophon uses the term *euzōnoi* even for hoplites (a heavy infantry unit), but they clearly fill a typical light infantry function. It is possible that these left some of their heavier gear with the baggage train to perform such 'light' tactical roles. Scott and Liddell suggested that these hoplites left their large *aspis* shields behind with the baggage train to increase their movement, speed and agility.[23]

Sekunda states that mercenaries during the Hellenistic period often acted as specialized light troops or *euzōnoi*.[24] Sparta's *euzōnoi* at the Battle of Sellasia (222 BC) were mercenaries.[25] At the Battle of Mantinea (208 BC), the majority of the Spartan and Achaean *euzōnoi* are also clearly defined as mercenaries by Polybius:

> 'At first the Tarentines alone were engaged, fighting gallantly, but as the light-armed infantry (*euzōnoi*) gradually came up to the support of those who were hard pressed, in quite a short time the mercenaries on both sides were mixed up.'[26]

That most of the *euzōnoi* at Mantinea were mercenaries is further emphasized by Polybius, who compares mercenaries under tyrants (in this case the *euzōnoi* fighting for Sparta) and those in service of a democracy (the *euzōnoi* serving in the Achaean League), claiming that those serving under a tyrant are superior to the mercenaries in a democracy's employ.[27] It is believed that these *euzōnoi* among the Hellenistic Greeks were armed as *thureophoroi* and *thōrakitai* (see Chapter 5: Thureophoroi and Thōrakitai).

Polybius, however, uses the term to describe any light infantry, from mercenaries to citizen soldiers, and even troops serving as subject-vassals. The word is used to describe light troops from a variety of cultures, whether they are Greek, Punic, Celtic, Roman or from further afield. For instance, Polybius refers to a variety of light troops at the Battle of Raphia (217 BC) in Seleucid employ as *euzōnoi*. The Agrianian and Persian archers and slingers, and a contingent of Thracians, are all described as *euzōnoi*.[28] The 5,000 Tralles/Trallians, Daae, Carmanians and Cilicians at the battle are also denoted as *euzōnoi*.[29] The Cilicians again appear in a Seleucid context as *euzōnoi* in the armies of Antiochus IV at the Daphne Parade.[30] At the Battle of the River Trebia (218 BC), Polybius refers to both the Carthaginian and Roman light troops as *euzōnoi*:

> 'When they were nearly at close quarters, the light-armed troops (*euzōnoi*) in the van of each army began the combat, and here the Romans laboured under many disadvantages, the efficiency of the Carthaginians being much superior.'[31]

Polybius is particularly fond of using *euzōnoi* to describe Roman light infantry, employing it more than any other term to refer to Roman light troops.[32] Polybius first used *euzōnoi* to describe Roman infantry in the first Punic War at the Battle of Panormus (251 BC) using their javelins to harass the enemy's elephants:

> 'When he saw that what he had designed was taking place he stationed some of his light troops (*euzōnoi*) before the wall and the trench, ordering them, if the elephants approached, not to spare their missiles, and when driven from their position, they were to take refuge in the trench and sallying from it again shoot at those elephants which charged at them.'[33]

Euzōnoi is undoubtedly one of several terms Polybius uses to refer to the *velites*, the javelin-armed citizen light infantry of the Roman manipular legion (see Chapter 5: Roman Velites/Grosphomachoi).

As mentioned above, *euzōnoi* is also used to denote light troops not only armed with range weapons, but those who could engage in a melee and carried some form of light armour. The *velites*, for instance, were primarily javelineers, but were more than capable of engaging in hand-to-hand combat. Polybius, describing the Battle of Mantinea (208 BC), groups *thōrakitai*, light troops wearing body armour and Illyrians among the *euzōnoi*.[34] *Thureophoroi* were also light troops under the banner of *euzōnoi* who could skirmish and fight in melee combat (for *thureophoroi* and *thōrakitai*, see Chapter 5: Thureophoroi and Thōrakitai).

When Polybius uses the word *euzōnoi* in a combat scenario, it never refers to the heavy infantry in disciplined ranks during pitched battles or static combat, but always to the tactical deployment associated with 'light' infantry. These duties comprise the following: scouting and surveying battle terrain and enemy forces (5.68.10, 18.19.9, 18.21.1); capturing high ground (3.101.5–7); forced marches (4.80.8, 5.61.8, 10.49.2); raiding and pillaging enemy territory (5.95.8); operating and fighting in difficult terrain (3.50.8, 4.14.6, 4.75.4, 5.68.11, 10.31.3–5); protecting a marching column (10.29.3–6); harassing the enemy's formation and positions (1.40.6, 3.102.2–3, 3.110.5, 4.11.6, 5.61.8); supporting heavy infantry (10.39.3, 18.23.7); operating alongside cavalry (2.27.6, 3.110.5, 5.95.8, 11.21.7, 11.22.5–8, 18.19.9); skirmishing before the main engagement of the heavy infantry (2.69.1–8, 3.73.1, 3.115.1–3, 5.52.7, 11.13.1–6, 11.21.7, 11.22.1–11); and outflanking an enemy position (3.74.2, 10.31.3–5).

From the above-mentioned examples, it becomes abundantly clear that Polybius prefers the term *euzōnoi* to denote light troops and that they operated in roles typically associated with such types.

Appian also uses *euzōnoi* to denote infantry operating in typical light troop roles. In his *Wars in Spain*, Appian writes that Gnaeus Scipio used *euzōnoi* during the Second Punic War to force-march to the aid of foragers who were attacked by Carthaginian troops.[35] In the *Mithridatic War*, Appian mentions that when the Pontic generals Neoptolemus and Archelaus first faced-off against the Bithynian troops at the start of the Battle of the River Amnias (89 BC), they only had their light troops (*euzōnoi*), chariots and cavalry ready, as the heavy troops of the phalanx were still on the march.[36] With these troops, they attempted to seize a hill in a desperate attempt to counter the numerical superiority of the Bithynian army.

Similarly, Diodorus Siculus uses *euzōnoi* to describe the light infantry of Antigonus in his preparations to campaign against the Nabateans. They fill a typical light infantry role, being equipped and fitted for speed to raid and pillage enemy territory:

> 'Deciding that this people [Nabateans] was hostile to his [Antigonus'] interests, he selected one of his friends, Athenaeus, gave him four thousand light foot-soldiers (*euzōnoi*) and six hundred horsemen fitted for speed, and ordered him to set upon the barbarians suddenly and cut off all their cattle as booty.'[37]

With the term *euzōnoi* translating as 'active', 'unencumbered', 'well-girdled' or 'well-equipped', and from interpreting the ancient literature, the *euzōnoi* were clearly troops equipped for roles typically associated with light infantry troops.

1.2.3 Gymnētēs

The term *gymnētēs* (γυμνῆτες) appears less frequently than *psiloi* and *euzōnoi.* It is, however, the preferred word used by Xenophon for light troops. *Gymnēs* translates to 'naked' or 'unarmoured' and is the origin of the word 'gymnast', as in ancient times Greeks trained naked at their local gymnasium. In military terms, it translates better to 'unarmoured'. It is, for instance, used by Herodotus to describe the Persian soldiers of Mardonius at the Battle of Platea (479 BC). Herodotus compares the Persian *gymnētēs* to heavily armoured Greek hoplites and attributes their defeat to the lack of armour:

> 'For what harmed them (the Persians) the most was the fact that they wore no armour over their clothes and fought, as it were, as naked/unarmoured men against men fully armed.'[38]

Euripides makes use of *gymnētēs* in the *Phoenician Women* to reference light infantry.[39] *Gymnētēs* is also used by Xenophon as a term for light infantry in the *Anabasis*, when the Greeks were asking for volunteers to capture a pass:

> 'Then they asked whether any one among the captains of light troops (*gymnētēs*) wanted to join in the march. The volunteer was Aristeas of Chios, who on many occasions proved himself valuable to the army for such services.'[40]

In the *Anabasis*, Xenophon also has the *gymnētēs* guarding the marching column (4.1.6), setting ambushes (4.6.17), fighting with range weapons (3.4.26, 5.2.12), occupying high ground (6.3.15) and performing scouting duties (6.3.15). In Xenophon's *Cyropaedia*, they are stationed on the vulnerable flanks of the heavy infantry.[41] In his *Hellenica*, the *gymnētēs* are described as very effective with their range weapons:

> '[T]he light troops (*gymnētēs*) of their adversaries did them a great deal of harm by throwing missiles and discharging arrows upon them from the heights upon their right.'[42]

Similarly, the Jewish historian Josephus, albeit writing in the first century, uses the term in his *Jewish Antiquities* (14.456) as well as in *The Jewish War* to refer to light troops armed with ranged weapons.[43] Appian, in his *Civil Wars* (*Bellum Civile*), also writing in the first century, describes the Roman army in Macedonia during the first century BC as containing a 'great number of archers and light troops (*gymnētēs*)'.[44]

Gymnētēs is undoubtedly a term used to describe light infantry. From reading the literature, the word is used as an umbrella term to describe light infantry and does not describe a specific type of soldier. Due to the term translating to 'naked' or 'unarmoured', we may assume that they carried very little or no armour.

1.2.4 Leves

Leves is most likely the closest Latin term used to refer to light troops in general. It literally translates to 'light', and in a military context is generally followed by the term *armaturae* and thus translates as 'lightly armoured'. In later Imperial Roman sources, the word *leves* is generally followed by *cohorts*. Because of the period under discussion, we will focus mostly on Livy. The term is used to describe both Greek and Roman light troops. As with *psiloi*, the word survived into the Middle Ages, where it was used by authors such as William of Tyre, describing the *Turcopoles* in the Frankish armies of the Second Crusade.[45] Here, however, the term no longer refers to infantry but light cavalry.

Livy describes the *leves* operating in roles typically associated with light troops. Livy refers to Macedonian light infantry (*levis armature*), who along with the cavalry secured water supplies for the exhausted heavy infantry of Philip V.[46] The cavalry and light infantry (*levi armatura*) then stood guard to protect the heavy infantry while they took rest. Again, we see Macedonian light infantry and cavalry inflicting heavy casualties among the Dardanians, due to the fact that the enemy had no light infantry or horsemen and that the difficult terrain aided the lightly armoured Macedonian troops while hindering the heavily armoured Dardanians.[47] The Macedonian *levi armatura* are also said to have been used to capture and occupy mountain passes during the Second Macedonian War.[48] At the Battle of Cynoscephalae (197 BC), the Roman *levi armatura* skirmished with the Macedonians before the main armies of the heavy infantry met for the melee engagement.[49]

Livy describes Philopoemen's Tarentine cavalry along with his Cretan infantry, whom he specifically calls *levis armature*, attacking the Spartan light troops.[50] At the Battle of Magnesia (190 BC), Livy refers to the Seleucid and the Pergamese Cretans and Tralles/Trallians as *levis armatura*.[51] The Pergamese *levis armatura* are specifically mentioned as fighting in a loose open order, as opposed to the close formation of the heavy infantry.[52] Pergamese *levis armturae* and cavalry were used by Attalus, the brother of Eumenes, to harass the numerically superior Seleucid forces invading their lands.[53] During Perseus of Macedon's war with Rome, the king's light troops (*levi armatura*) and cavalry are used to form the rearguard to protect the marching column.[54] Perseus' light troops were also used to capture and hold strategic mountain passes against the Romans.[55] During the Second Punic War, Fabius used light troops (*levi armatura*) and cavalry to guard foragers.[56] Livy also highlights Philopoemen's concerns regarding adequate light troops (*levis armturae*) for protection when marching his heavy troops through difficult terrain.[57]

At the Battle of Mount Olympus in Anatolia (189 BC), fought between the Roman Republic and the Galatian tribe of the Tolostobogii, the light troops played an overwhelmingly dominant role in achieving victory for Rome (see 7.3.1, The Battle Mt Olympus in Anatolia).[58] During the campaign, light troops were specifically used in this mountainous and difficult terrain to drive the enemy from the hills and high ground. Livy groups the Cretan archers and slingers supplied by King Eumenes of Pergamum, the Tralles/Trallians and Thracians, as well as the Roman *velites*, as light troops (*levi armatura*).[59] These light troops, armed with typical range weapons, gave an excellent account of themselves, as their 'arrows, sling-bullets, darts, coming from all sides wounded them (Galatians) unexpectedly … struck by light weapons, coming from unseen and distant sources'.[60] *Levis armaturae* is undoubtedly one of the terms Livy used when he referred to the Roman *velites*, and he often used the terms interchangeably (see Chapter 5: Roman Velites/Grosphomachoi). Livy himself states that the *velites* were the light troops of the Roman Army during the middle Republic: 'The *velites*, the light infantry (*levis armaturae*) of those days, were stationed at the head of the lanes between the columns of maniples.'[61] When Livy refers to *levis armaturae* within the Roman Army, he is more often than not referring to the *velites*.

The word *leves* is also used in late Republican sources to refer to light troops. By the time of the late Republic, the *velites* were no longer a troop type in military deployment and the term now refers to foreign auxiliaries, but it could also mean lightly equipped legionaries. It is, as in Livy, also used to refer to light troops of Rome's enemies.

In his *Gallic Wars* (*De Bello Gallico*), Caesar used his slingers, archers and Numidian infantry, who are referred to as '*levis armaturae*', to quickly reinforce a river crossing against the attacking Belgae.[62] *Levis armaturae* were again force-marched to come to the assistance of hard-pressed Roman cavalry fighting Gallic tribesmen.[63] We also see Caesar's cavalry operating alongside archers and slingers, whom Caesar refers to as '*levis armaturae*'.[64] Caesar also conscripted horsemen and light infantry (*levis armaturae*), 'who were accustomed to fight alongside them', from Germania.[65] According to Caesar, the Gallic tribes at the Battle of Alesia (52 BC) used this same tactic of intermixing their cavalry with light infantry to 'sustain the impetuosity of our [Roman] cavalry' and successfully held off a Roman attack against the retreating Gallic troops.[66] Furthermore, Caesar made use of cavalry and light troops to protect his vulnerable foragers from Gallic attacks.[67]

In Caesar's *Civil Wars* (*De Bello Civili*), the Lusitanian *levis armaturae* and Spanish *ceatrati* constantly ambushed and attacked Caesar's foragers.[68] As in the *Gallic Wars*, we see Germanic *levis armaturae* operating alongside cavalry, patrolling river crossings.[69]

Interestingly, Caesar refers to Pompey's archers and slingers attacking one of his fortified positions, stating that these archers and slingers were supported by *levis armaturae*.[70] Since the *levis armaturae* were mentioned separately from the archers and slingers, it is possible that they might have simply been reinforcements of lightly equipped legionaries, or perhaps an additional group of light troops with mixed equipment. There are several passages where Caesar mentions 'archers and light troops (*levis armaturae*)', but this should in no means be seen as a distinction to say archers were not light troops and should rather be read as 'archers and other light troops'.[71]

The term *levis armaturae* also appears in late Republican writings such as Cicero, who refers to the three main branches of the army: the cavalry, heavy infantry and the light troops (*levis armatura*).[72] In his *Concerning Divination* (*De Divinatione*), he again refers to *levis armaturae* and connects their main purpose to skirmishing.[73] In this particular case, the term is used as a metaphor to denote a quick reply or rebuff.

Frontinus, writing in the first century AD, mentions Spanish '*levis armaturae*' during the late Republican era, referring to them as being masters of ambuscades.[74] The term also appears in the writings of second-century AD writers such as Suetonius and Tacitus.[75] While Tacitus wrote on Imperial Roman history, it is to be noted that imperial '*levis*' filled roles similar to their earlier Republican counterparts; they operated alongside cavalry and raided enemy lands, fought with ranged weapons and deployed to the front and in the flanks of the heavy infantry in pitched battles.[76]

Even though far beyond the period discussed in this book, it is worth mentioning that the term *levis armaturae* appears in Roman military writing as late as the fourth century AD in works such as Ammianus Marcellinus' *Res Gestae*. Ammianus was a military officer during the reign of Constantius II and Julian the Apostate who later became a historian. In the *Res Gestae*, Ammianus refers to Roman light infantry in the army of Julian the Apostate as '*levis armaturae auxiliis*'.[77] On this particular occasion, these troops accompanied the emperor himself to various points of the battleline. Because of their light equipment, they could rapidly reinforce any point which was hard-pressed.

1.3 Light Infantry as a Collective Term

From the above assessment, it is clear that military historiographers made use of common terms to refer to light troops in general. These various collective terms were used across a broad time period and employed by contemporary writers, from those who were accomplished military officers – such as Xenophon, Polybius and Caesar – to philosophers and historians who were armchair generals at best, including Appian and Livy. The authors who used these terms understood, or at least expected their audience to comprehend, what these terms meant, so much so that they were even used, as we saw in the case of Cicero, as a metaphor.

As we have seen, these troops operated outside the static close formation of pitched battles. These collective terms focused on their tactical deployment and function, rather than being based on their armament alone. They could thus refer to anything from soldiers who wore no armour at all and avoided melee engagements at all costs, to troop types clad in various forms of 'lighter' armour and equipped to skirmish and engage in melees. These collective terms were also not ethnic or class-specific, and could refer to the poorest of citizens, foreign mercenaries, conscripted levies from the far-flung corners of the various empires of the time, allied and enemy forces and even wealthy or well-to-do citizens who donned lighter war gear so that they could operate in the roles of light troops. Some of these troops were highly sought after for their skill and were employed across the ancient world as elite troops.

Part I: Light Infantry Troop Types

This chapter has focused on collective or umbrella terms describing light troops in general. The following section is more specific and focuses on the various individual troop types and the Greek and Latin terminology used by the ancient authors to describe them. It is a brief assessment of the period sources and explores the different arms and armour used by the light infantry, as well as assessing their tactical deployment. Chapter 2 examines archers of the Graeco-Roman world. Our sources are predominantly Greek; while archers did serve in Republican Roman armies, they played a far smaller role than in other armies from the period, serving as mercenaries or troops from vassals or Roman allies, generally Numidians or Cretans.

Chapter 2

Archers: Toxotai/Sagittarii

Archers

The Greek term *toxotai* (τοξόται) and Latin *sagittarii* translate as bowmen or archers. As their name implies, they were range units armed with bows. Interestingly, our modern word 'toxic' has its origins in the Greek *toxon*, meaning bow, presumably from poison arrows shot from the weapon; images of Heracles shooting poisoned arrows (having dipped them into the blood of the Hydra) at the Stymphalian birds come to mind. Archers were deployed on countless battlefields throughout history. Indeed, the bow, as a weapon of war and violence, far predates the Egyptian pyramids and Britain's Stonehenge. Ötzi the Iceman, who died more than 5,300 years ago during the late Neolithic period, met a violent and bloody end in the Ötztal Alps, with a flint-tipped arrow lodged in the left shoulder of his mummified remains.[1] Ötzi was surely not the first to have his life snatched away by an arrow, with evidence suggesting that the bow was used in armed conflict as far back as 10,000 BC.[2]

In the ancient Near East, the bow was always held in the highest regard. It often served as a symbol of royal strength in the great imperial kingdoms of Egypt, Assyria and Achaemenid Persia.[3] It was undoubtedly the bow, and the Parthians who wielded it with exceptional skill, that brought a halt to Rome's expansion eastwards at the Battle of Carrhae (53 BC). Under the Sassanian Empire, the bow remained the royal weapon of the king, as several Sassanian iconographical sources demonstrate.[4] With the rise of Islam, archery was quickly absorbed into the religion, and the fourteenth-century manual *Saracen Archery* proclaimed that the practice of archery should be a religious obligation.[5] The sentiment is echoed in the sixteenth-century manual *Arab Archery*.[6] Throughout the Ottoman Empire, the bow and arrow held a sacred position, with archery practised in every city, town and village.[7]

Among the Greek and Roman cultures, however, archery was viewed quite differently. Within the period's literature, the roles of archers are quite often glanced over at best, and at times it seems that archery is held in contempt as a martial skill. As far back as Homer's *Iliad*, archers are regarded as less heroic than their shield-bearing counterparts.

In the *Iliad*, on both sides in the Trojan War, legendary heroes seem to look upon archers with scorn. Diomedes, after being wounded by an arrow shot from the bow of Priam's son Paris, addresses Paris as an 'archer' (*toxotēs*), using the term as an insult.[8] Diomedes then compares the 'blunt dart' of Paris to a blow delivered by a woman or child, while boldly claiming that the slightest touch of his own spear makes widows of women.[9] The Trojan hero Achamas degrades the Greeks by calling them 'arrow-fighters' (*iomōroi*).[10] This low regard for archers continued in Classical Greek literature. Fifth-century BC playwriter Sophocles, in his *Ajax*, has Menelaus look down on Teucer because he is an archer and not a shield-carrying hoplite.[11] Euripides, in his *Heracles*, in similar fashion, has Lycus question Heracles' ability and courage (*aretē*) in battle because he is not a heavy infantry soldier armed with a shield and spear, but an archer carrying a bow, the weapon of a coward:

> '[Heracles] never buckled shield to arm nor faced the spear, but with a bow, that coward's weapon, was ever ready to run away. Archery is no test of manly bravery; no!'[12]

While Lycus is the play's villain and the opponent of the hero Heracles, there is enough other literature that portrays archery in a negative light. Both Kelly and Pritchard highlight the dichotomy used by Aeschylus in his *Persians*, which goes to great lengths to portray the Persians as cowardly archers and the Greeks as brave, spear-wielding hoplites.[13]

In Herodotus' *Histories*, a Spartan hoplite named Callicrates is mortally wounded.[14] He is dismayed that he will not die an honourable death from another warrior's spear thrust, instead losing his life to a loathsome archer. In Thucydides' *The Peloponnesian War*, one of the Spartan prisoners from the Battle of Sphacteria (425 BC) attempts to effeminise archery by referring to the arrows that brought the Spartan defeat as 'spindles' from a woman's spinning wheel.[15] This same man further remarks that arrows would be worth something if they could pick out the cowardly from the brave. Both these cases highlight the dishonour of dying at the hand of an archer and his bow. Thucydides' account of light infantry combat (which obviously includes archery) during the Peloponnesian War suggests that the archers' arms and fighting style are largely inconsequential:

> 'First, the stone-throwers, slingers, and archers of either army began skirmishing, and routed or were routed by one another, as might be expected between light troops; next, soothsayers brought forward the usual victims, and trumpeters urged on the heavy infantry to the charge.'[16]

While the Roman sources are less hostile towards archery, they are far from singing its praises. Throughout the Republican period, archers or *sagittarii* feature almost as a side note. During this period, Roman citizens do not appear to have served as archers at all. As mentioned previously, they are always foreign mercenaries or allied troops, especially in the form of Cretans. While archers served in Roman armies throughout the Republican period, they are only mentioned in passing, their martial prowess and contributions in battle almost never mentioned. During the later Empire and Byzantine periods, as Rome faced ever-increasing pressure from the East from cultures such as the Huns, Avars and Sassanids – who primarily focused on archery (particularly horse archery) – there was a resurgence in the importance of the bow as a weapon of war for Roman forces.

While the literature does not glorify archery, and in many ways treats the martial skill with some contempt, the brutal realities of the ancient battlefield paint a very different picture. In practice, the value and effectiveness of archery were understood. One does not have to read too deep between the lines to see that generals realized the need for archers and their skill of killing at a distance. After all, the arrow brought death and could lay low the loftiest and most skilled of warriors.

While Diomedes' above-mentioned words attempt to brush off the arrow and the bow as inconsequential, the severe wound inflicted upon him by Paris tells a different story. Diomedes is unable to leave the battlefield on his own strength, and it is up to Odysseus, his comrade in arms, to defend him from the enemy onslaught until the wounded hero is taken to safety in a chariot.[17] One must also not forget that even the great Achilles, the most martial warrior of the *Iliad*, was sent down to Hades after being struck by an arrow.

In his *The Persian Wars*, Herodotus is silent on whether archers fought on the side of the Greeks at Marathon (490 BC) and Thermopylae (480 BC). At Platea (479 BC), however, there is a contingent of Athenian archers who gave a good account of themselves, seeing off a Persian cavalry charge. A well-placed arrow brought down the horse of the Persian cavalry commander, Masistius.[18] Having been thrown from his mount, he quickly succumbed to a hoplite's spear. The value of archers seems to have been recognized by Pausanias, the Spartan regent and general at the Battle of Platea. Hard pressed by the Persians, Pausanias specifically requested Athens to send archers to the Lacedaemon (Spartan) position when the Persian cavalry threatened to overrun their lines.[19]

By the time of the Peloponnesian War (431–405 BC), Athens already had a standing corps of 1,600 archers.[20] Athens, at the head of its fledgling empire, could afford to invest in military expenditure, and there is sufficient evidence to suggest that these archers enjoyed full-time employment.[21] According to Thucydides, the archers were paid one drachma a day.[22] Pritchard calculated that the annual cost of the 1,600 archers was 96 talents, approximately 10 per cent of Athens' annual income.[23] For Athens to have devoted so much of its financial resources to bowmen speaks volumes for the value that archers brought to the battlefield. Membership of the archer corps consisted of both citizens and *metics* (tax-paying foreigners living in Athens).[24] A fifth-century BC inscription, for instance, sets out the tax on the land forces of Athens, stipulating three *obols* for the citizen (*astoi*) and foreign (*xenoi*) archers.[25] The fact that the Athenian archer corps had *toxarchoi* (commanders of the archers) demonstrates that they were well organized.[26] Archers are mentioned numerous times throughout Thucydides' *The Peloponnesian War*, and while the accounts certainly favoured the heavy infantry hoplites, it is clear that light troops, particularly archers fighting on the side of Athens, played an important role throughout the war.[27] The importance of archers during the period is further highlighted in Sparta's realization that its hoplites alone could not win this war, and that archers and cavalry were needed:

> 'Since Pylos was in enemy possession as well as Cythera and they [the Spartans] were beset on every side by a war that required mobility and against which it was impossible to take precautions, the Spartans contrary to their usual custom established a force of four hundred cavalry as well as archers.'[28]

The Athenian standing archer corps was a novelty among the Greek states, with other states relying primarily on mercenaries to fill their light infantry ranks with archers. Sparta, for example, often hired Cretan bowmen, a tradition which possibly dates back to the Messenian Wars of the eighth and seventh centuries BC, if Pausanias is to be believed.[29] The islanders from Lesbos, during their revolt against Athens, relied on mercenary archers from the Black Sea, possibly Scythians.[30] Andocides, in his *Oration*, also mentions that Athens hired a mercenary force of 300 Scythians. In addition to their own archer force of 400 men, Athens hired eighty Cretan bowmen for its Sicilian expedition in 415 BC.[31] Cretans serving as mercenary archers were highly sought after, serving across the Mediterranean world in Greek and Roman armies. Pausanias mentions that the Greeks were unaccustomed to using the bow as a weapon, but that the Cretans were the exception.[32] A section below is specifically dedicated to Cretan mercenary archers (see Chapter 6: Cretan Mercenaries).

While Xenophon does not specifically mention Greek archers at the Battle of Cunaxa (401 BC), it is possible that there were bowmen grouped among the 2,500 light troops at the battle. There are several accounts in the *Anabasis* referencing Greek archers on the perilous journey back to Greece. Xenophon, for instance, mentions a specific number of Cretans archers, 200 in total, who accompanied the exiled Spartan general Clearchus in support of Cyrus the Younger.[33] Whenever Xenophon refers to archers on the Greek side they seem to be Cretans, and it is possible that the Greek archers of the *Anabasis* consisted entirely of Cretans. The main purpose of the Greek forces hired by Cyrus the Younger for his rebellion against his brother Artaxerxes II was to supply Cyrus with heavy infantry. Persia always enjoyed a martial culture focused on archery, and thus did not need to hire more archers, hence the small group of Cretans in the *Anabasis*. These Cretans did, however, give a good account of themselves on several occasions throughout the long march home. The Greeks' eastern adversaries enjoyed a marked numerical superiority over the Greeks when it came to archery, Mithridates on one occasion attacking the marching Greeks with 4,000 archers and slingers.[34] The Greeks suffered severely under a hail of arrows and stones shot from such a large force:

> 'They not only inflicted many wounds, but they got the better of the Greek light troops and shut them up within the lines of the hoplites, so that these troops, being mingled with the non-combatants, were entirely useless throughout that day, slingers and bowmen alike.'[35]

When the Greeks finally escaped the Persian attack, they took shelter in a small village to treat the wounded: 'There they appointed eight surgeons, for the wounded were many.'[36]

According to Xenophon, the Persian archers were also able to outrange the Cretans.[37] This, however, was not due to the Cretans being less skilled than their Persian adversaries or the Cretans using inferior bows, but rather due to using different arrow types. It was not only Persian archers that posed a threat to the Greeks. On their return route, the Greeks were attacked by the Carduchians, whom Xenophon and Diodorus claimed used large bows, 3 cubits (1.38m) in length, and arrows 2 cubits (0.92m) long that could pierce through the shields and armour of the Greeks.[38] With these, the Carduchians killed and wounded many of the Greek soldiers. The arrows were so large that the Greeks used them as javelins, throwing them back at the enemy with *ankylē* throwing straps (see Chapter 8: Javelin and Throwing Strap).

The literature during the first half of the fourth century BC is fairly tacit about archery. We do, however, see that Agelaus II, king of Sparta, has archers in his army collected at Ephesus in 395 BC during Sparta's brief hegemony period.[39] Agelaus had his army perform their skills in the hippodrome of the city and offered prizes to those who were 'most skilled in their respective duties', including the *peltastai* and *toxotai* 'practising their shots'. Unfortunately, the sources fall silent on Agelaus' archers fighting the Persians.

They do appear once again upon Agelaus' recall to Greece to face the hostile coalition of Corinth, Athens and Thebes in 394 BC.[40] During this conflict, Corinth had a contingent of archers who occupied the broken hilly terrain and rained arrows down on the Spartans' 'unshielded' flank, 'and did them a great deal of harm by throwing missiles and discharging arrows upon them from the heights upon their right'.[41] Sparta once again made use of Cretan archers, 300 in number, against the coalition forces at the Battle of Nemea (394 BC), and an unspecified number of Cretans served in Agesipolis' expedition against the Argives in 388 BC.[42]

Macedon's rise under Philip II and Alexander the Great saw the newly innovated *sarissa*-bearing heavy infantry of the Macedonian phalanx dominate the battlefield, but archers still featured extensively throughout Philip's and particularly in Alexander's campaigns. There is evidence that Philip and Alexander had a contingent of Macedonian archers.[43] Arrian, for instance, refers to a contingent of 'Macedonian Archers' stationed on the right wing at the Battle of Gaugamela (331 BC), while another contingent of archers (ethnicity unspecified) deployed with half the Agrianians and Companions (*hetairoi*) on the left.[44] Sekunda states that it was possible that Philip and Alexander, unable to rely solely on Cretans, created the corps of Macedonian archers due to the need for bowmen during this period.[45] If this was indeed the case, the Macedonian contingent of archers would be the only Macedonian infantry unit outside of the Macedonian phalanx. Unsurprisingly, the Cretan mercenary archers feature in the armies of Philip and Alexander.[46] After Gaugamela, the recruitment pool of archers expanded dramatically as eastern bowmen, including horse archers, became available to Alexander and his Successors after the conquest of the Achaemenid Empire.

During the Hellenistic era, due to the immense size of Alexander's Successor states, archers from a variety of ethnicities and cultures were fielded and their numbers were dramatically increased. For instance, Peukestas, satrap of Persia, who fought alongside Eumenes at the Battle of Gabiene (316 BC), had an army of 10,000 archers and slingers.[47] Pyrrhus of Epirus moved an immense army across the Ionian Sea, which included a force of 2,000 bowmen (ethnicity not mentioned), 10 per cent of his entire infantry force.[48] The Seleucids, who inherited the largest portion of Alexander's eastern conquests, were able to field copious numbers of bowmen; we hear of Persian, Elymaean and Mysian archers numbering in their thousands.[49] Cretan archers once again feature extensively throughout the Hellenistic period in Antigonid, Ptolemaic and Seleucid armies, as well as among the forces of smaller Greek states.

Further west, archers served in the armies of Magna Grecia and the Roman Republic. The Syracusan tyrant Gelon, according to Herodotus, offered to send 2,000 archers to aid the Greeks against the Persians,[50] while Hiero II of Syracuse provided his Roman ally with 1,000 archers and slingers during the Second Punic War, specifically to counter the Carthaginian light infantry.[51] Archers, generally of Cretan and Numidian stock, appear in Roman armies during the Pyrrhic (Dion. Hal., *Ant. Rom.*, 20.1–2), Punic (Livy 22.37.8), Syrian (Livy 37.41.10–12), Galatian (Livy 38.21.1–4), Macedonian (Livy 42.35.6), Gallic (Caes., *Gal.*, 2.10.1) and Civil Wars (Caes., *Civ.*, 3.45.2–4).

All this indicates that archery was an intricate part of Greek and Roman warfare, with bowmen fielded on almost every battlefield of the ancient world. While the literature might not heap glory and praise upon the bow and arrow, archery was clearly an important feature of combat. But if archery featured so prominently in armed conflict, why the prejudice towards bowmen? Did the Greeks and Roman really view archery with contempt? The answer is not so simple, having more to do with the status of the archer than with his skill with bow and arrow.

Archers, along with most other range troops, simply did not fit into the ethos of Greek *arête* or Roman *veritas*. For the Greeks and Romans, manly excellence and virtue was meeting the enemy in hand-to-hand combat. To them, the ideal concept of courage and bravery demanded that a warrior 'stand there, shoulder to shoulder, with no fear of death, and fight for their country'.[52] The heavy infantry, such as Greek hoplites and Roman legions, embodied the ideal soldier, standing their ground in disciplined ranks against the onslaught of the foe, risking their personal safety and even sacrificing their lives for their country and their honour. This is what immortalized the 300 Spartans at Thermopylae and the last stand of the Roman hero Horatius against the Etruscans. This was not the fighting style of archers, or light infantry in general, who fought from afar and did not stand their ground against the enemy but ran from the spear. For the heavy infantry, giving ground to the enemy was seen as disgraceful. Brasidas, the distinguished Spartan officer during the Peloponnesian War, highlighted the lack of shame and manliness of the Illyrians, who were willing to abandon ground when hard-pressed.[53] During the same war, Corinthian hoplites, after suffering a defeat by the Athenians, were shamed to such an extent by the elders of their city upon their return that they marched back to fight another battle against those who had just bested them, only to be defeated once again.[54] As far back as the *Iliad*, we are told that the Locrians were armed with bows and slings because they lacked the courage and heart to fight in close combat.[55]

The Roman military discipline of *decimatio* (the origin of the modern term 'decimation') was a punishment for those who gave ground to the enemy or showed cowardice on the field of battle, with the execution of every tenth man of their group:

> '[I]f, for instance, some entire maniples have quitted their ground in the presence of the enemy, it is deemed impossible to subject all to the *fustuarium* or to military execution ... The Tribune assembles the legion, calls the defaulters to the front, and, after administering a sharp rebuke, selects five or eight or twenty out of them by lot, so that those selected should be about a tenth of those who have been guilty of the act of cowardice.'[56]

A soldier was obviously expected to stand his ground against the enemy, but the archers, lacking the arms and armour for melee engagements, had no choice but to quit the field when the enemy charged at them and thus could not live up to the ideal warrior ethos of the Greeks and Romans.

Certain war gear had certain social implications among the Greeks and Romans. Since the acquisition and maintenance of horses was expensive, only the nobility and higher classes of Greek and Roman society filled the ranks of the cavalry. In Classical Greece and Republican Rome (prior to the Marian Reforms of the late second century BC), the heavy infantry had to supply their own arms and armour, restricting the heavy infantry to the middle classes who could afford the expensive equipment. This left the poorest of citizens and foreigners (*xenoi*) to fill the ranks of archers and other light troops. The standing corps of archers was definitely considered the poorest of the land forces prepared by Pericles for the war against Sparta. These archers only paid half the tax for the upkeep of the Lyceum compared to their hoplite counterparts.[57] Throughout history, the lower classes and poor were always looked down upon. Prior to the reforms of Marius, a Roman citizen could not even join the military if he did not own a certain amount of property, and foreigners could only join as mercenaries or conscripted allies. The poor were simply deemed not worthy of fighting because they did not have property to fight for or to defend.

I have already mentioned that the Greek states, the Successor empires and Rome largely relied on mercenaries, vassals and auxiliaries to fill their archer needs. Mercenaries have always held a dubious reputation of disrepute. Isocrates, for instance, calls them 'the common enemy of all mankind'.[58] This is understandable, as these men's loyalty lay not with their homeland, their citizenship or the cause of the war, but rather to the coin that brought them to the battlefield. They did not face the enemy and the dangers of battle out of a sense of civic duty, but purely for financial gain. If they died, the coin could not be spent, and since the cause of the war was not theirs, their death – at least from their own perspective – would be in vain. Their loyalty was thus always questionable. Furthermore, most men who became mercenaries were those who found themselves in financially desperate situations and were forced to find a means of income through becoming soldiers of fortune. Many mercenaries were therefore from the lower strata of society, further tainting their image.

It was thus not the case that the Greeks and Romans looked down on archery or deemed it an inferior or useless skill. On the contrary, the advantage that archery brought to the battlefield was highly valued. Archery would not have featured so extensively across such a broad period, and the demand for specialists such as the Cretans would not have been as high as it was, if the art of archery was seen as inferior. While archery itself was not looked down upon by the Greeks and the Romans, the bowmen were, simply due to their social and economic status, being outsiders, foreigners and the lowly poor.

2.1 Arms and Armour

2.1.1 The Bow

The archers in Greek and Roman armies made use of the composite bow. This type of bow comprised three different materials: horn, sinew and wood. The self-bow, meanwhile, was a single stave of wood, such as the well-known English longbow of the Middle Ages. The composite bow is in essence a recurve bow, meaning the limbs of the bow bend away from the archer when unstrung (**see plate 1**). These bows are creative and of complex design, with the various materials shaped and glued together in a lengthy process. The wood, which would be non-resinous to allow for the glue to take well, formed the core of the bow. The sinew was glued onto the back of the bow (the part facing away from the archer when drawing the bow), while the horn was glued onto the belly (the part facing towards the archer when drawing the bow) (**see plate 2**). The glue which held everything together was usually made from boiled fish parts or boiled hide. The combination allowed for a balance of tensile and compressive strength. When the bow is drawn, the sinew stretches while the horn compresses, allowing the bow limbs to store an immense amount of energy (**see plate 3**). This energy is transferred to the arrow upon release, allowing the arrow to be shot with greater force and speed. The art, craftmanship and dedication involved in making a horn bow can only truly be appreciated by experiencing the process itself. Csaba Grózer from Grózer Archery, one of the top traditional bow makers, posted a video where he takes the viewer through every step of traditional composite bow making, explaining and demonstrating the intricacies and complexities involved.[59]

The advantages which composite bows held over self-bows was immense. The composition of materials made them far more resilient to the strains of drawing the bow in comparison to self-bows. This in turn allowed composite bows to be of a small yet powerful nature, whereas self-bows had to be far greater in length and girth to generate the same amount of power.

There are some setbacks when it comes to such composites bows. They were, for example, very susceptible to the climate. Extreme temperatures and especially wet conditions could easily affect the glue holding the various parts together, thereby compromising the integrity of the weapon. This could lead to distortion or delamination, and the bow needed constant expert care. This is probably best described in the sixteenth-century archery manual *Kitāb ghunyat at-tullāb fī ma'rifat ramy an-nushshāb* ('Essential Archery for Beginners'), which states:

> '[A]n archer should never neglect his bow for a single moment, and in extremes of temperature he should inspect it day and night, hour by hour, and not let it out of his mind even if he is sure that it is stable and true. ... when the weather is cold, his best policy is to put the bow inside his clothes and warm it with his body ... When going to bed at night, he should also keep the bow inside his clothes to protect it against the damp.'[60]

The composite bow dates back to at least third millennium BC Mesopotamia.[61] In a Greek context, the Homeric epics refer to composite bows, such as Pandaros' 'polished bow of the horn of a wild ibex', Dolon's 'back-strung bow' and the bow Odysseus used against Penelope's suitors, also constructed of 'horn' and 'back-strung'.[62]

During the early sixth century BC, numerous images of 'Scythians' start to appear in Greek vase painting (**see plate 4**). Whether these were actual ethnic representations of Scythians or they were meant to symbolize second-class characters, or even simply represent Greeks in the archetype of the epitome of great archers (the Scythians being excellent bowmen), is not yet conclusive. There is, however, strong evidence for the latter two arguments.[63] Nevertheless, the bows which feature in the vase paintings are undoubtedly of the composite type, as made evident by their dramatic curves and recurved ends standing above the bowstring when braced. Literature also supports its composite construction. Aeschylus, for instance, refers to the Scythian bow as 'back-bent'.[64] Scythian bows have a fairly unique double convex shape, and in the words of Mike Loades, 'No other bow embodies both extremes of reflex and deflex to quite the same extent as the Scythian bow' (**see plates 1–4**).[65]

Unfortunately, because of the biodegradable materials of composite bows, archaeological bow finds are extremely rare, so we have to rely on iconographic evidence and the few descriptions of bows in literature. However, excavations of graves at the Yanghai cemetery in Xinjiang, north-west China, produced several partial bow finds, as well as an almost intact 3,000-year-old bow of the Scythian type.[66] The frozen arid desert climate of Xinjiang contributed to it being exceptionally well preserved. The resemblance of the preserved bow to that of the Scythian bows in Greek art is uncanny. The Yanghai find is asymmetrical, with the top limb of the bow a little longer than the bottom limb. This is also reflected in some of the Greek vase paintings, the image of Odysseus slaying the suitors serving as an example.[67] The fact that the particular bow design stretched from Greece all the way to China speaks very highly of its effectiveness.

In 2009, Adam Karpowicz reconstructed a working replica of the Yanghai bow.[68] In modern terms, the power of a bow is calculated in 'draw weight', and even the most novice of archers has a fair understanding of a bow's capabilities when they are given the bow's draw weight in pounds (lb). The replica Yanghai bow had a draw weight of a staggering 54kg (120lb) at a modest draw length of 71cm (28in). Arrows which were found alongside the bow measured 76–79cm (30–31in); longer draw lengths were thus possible, which would dramatically increase the draw weight. While the exact draw weight of the original is impossible to determine, Karpowicz gave it an estimate of between 36 and 63.5kg (80–140lb). Even at its lowest estimation of 36kg (80lb), this was an extraordinarily powerful bow, especially for its size of approximately 112 cm. To put this in perspective, modern mechanical compound bows are maxed out by manufacturers at 36kg (80lb). In order to draw a modern compound bow, the frame is fitted with a complex pulley system to assist the archer. Traditional bows have none of these mechanical aids, and to draw such powerful examples the archer has to rely on years of training and experience. It is thus only a very experienced and capable archer who is able to draw a traditional bow of such draw weight. Scythian bows were clearly powerful weapons. The prominence of the Scythian-style bows in Greek art makes it clear that Greek archery was highly influenced by the Scythians. There seems little doubt that the Greeks adopted the Scythian bow, and we find much iconographical evidence of Greek gods, goddesses and heroes, and the occasional Greek soldier, armed with Scythian-style bows.

Greece's tumultuous relationship with Achaemenid Persia brought it into contact with Persian archers and the Persian composite bow. Unlike the Scythian bow with its distinctive double convex shape, the Persian bow was of a single convex nature when strung (**see plate 5**). No bow from the Achaemenid period survived the ravages of time, but there are numerous iconographical representations of Achaemenid Persian bows in art. None is more famous than the Frieze of Archers from the Palace at Susa, which is believed to portray the Persian Immortals, the royal guard of the Persian kings **(see plate 6)**. The Achaemenid Persian bow was in all likelihood a shorter adaptation of the Elamite bow.[69] Elamite-style bow tips or ears (*siyah*) often end in the iconic shape of a duck's head.[70] Herodotus describes the Persian bow as 'great (*megála*/μεγάλᾱ)'.[71] This should be interpreted as great in strength rather than size, as the majority of Persian bows depicted in iconography are of a moderate size, albeit bigger than the Scythian bow.

The Persian Achaemenid bow, even though it is by far the most represented in Persian iconography, was not the only bow used by the Persians. The Scythian bow design, which was used across the vast steppes from the fringes of Europe to central Asia, had made its way into Mesopotamia and was adopted by the peoples of Iran.[72] Herodotus does mention that Persia hired Scythians to teach future Persian troops their skill in archery.[73] While Herodotus is not the most trustworthy of sources, there are a few cases of double reflex Scythian-style bows in Persian iconography. Persian kings fielded Scythians on numerous battlefields, from Marathon (490 BC) to Gaugamela (331 BC), and the Achaemenid rulers even waged several campaigns against the Scythians, with mixed success. The double reflex bow adopted by the Persians has less-extreme angular curves than the earlier Scythian bow. This more elegant curved 'B'-shaped bow appears in both Greek and Persian iconography, and can be seen wielded by Persians, Scythians and Greeks. The fifth-century BC Tatarli Tomb mural, for instance, depicts a battle between Scythians and Persians. The Persian infantry are armed with the larger single reflex Elamite bow, while the horsemen on both the Persian and Scythian sides are armed with smaller double reflex bows. The imperial guardsmen portrayed on bas-relief at Persepolis also carry two types of bows: the Persians carry a single reflex Elamite Persian bow slung over their shoulders, while the Medes carry a shorter Scythian-style double reflex bow in their *gorytos* (a combined quiver and bow case, also of Scythian origin) **(see plate 7)**. The Greek-manufactured Kul-Oba kurgan vase of the fourth century BC serves as another example of Scythians using the 'B'-shaped bow.

As portrayed in the Tatarli Tomb, it is very likely that the smaller, more manoeuvrable bows were preferred by the cavalry, while the larger bows were used by the infantry. It is generally understood that horse archers used lighter draw-weight bows than their foot counterparts.[74] Because infantry have their feet planted securely on the ground, they are able to make use of their whole body to draw the bow, allowing them to draw larger and more powerful bows. In contrast, a horse archer only has his upper body with which to draw the bow. This is reflected in the writings of the eighteenth-century Chinese emperor, Yongzheng, who states that bows with a draw weight of 80lb are sufficient for horse archers:

> 'If there are those who wish to learn how to use a hard bow, they should practice naturally, gradually increasing the strength of the bow ... Besides, using a hard bow on horseback is difficult, so what is the advantage? A bow that is of strength six [80lb] or greater is enough.'[75]

Shorter bows such as the 'B'-shaped Scythian bow were, however, not exclusively used by horse archers, as there are depictions of Scythian, Persian and Greek infantry wielding them. Smaller bows were in all likelihood preferred by many infantry archers, such as the Cretan mercenaries, serving in Greek and Roman armies because they operated in difficult rough terrain (see Chapter 6.3: Cretans at War).

While there is very little iconographical evidence for Parthian archers, it comes as no surprise that the Scythian-style double reflex bow was their weapon of choice, since the Parthians themselves originated from the Scythian steppes. Depiction of various Parthian rulers on silver drachms shows both types of Scythian style bows, those with extreme curves and the more elegant and rounded versions.[76] The power of the Scythian bow, which Plutarch describes as 'large and mighty and curved so as to discharge their missiles with great force', is perhaps best illustrated in his account of the Battle of Carrhae (53 BC), where the Romans suffered a terrible defeat at the hands of the Parthians:

> '[The soldier] showed him [Publius] that their hands were riveted to their shields and their feet nailed through and through to the ground, so that they were helpless either for flight or for self-defence.'[77]

The Scythian-style bow remained popular throughout the Graeco-Roman world and beyond, and it was only during the third century AD that the design of the double reflex bow saw changes to its design. The double reflex 'B' shape was retained, but the *siyahs* or ears were extended, which in turn allowed for a smoother, longer draw of a more powerful bow.

Range of Ancient Bows

It is difficult to determine the range of ancient bows, there being several factors which determine how far an arrow could be shot. The bow's shape and construction, draw weight, the material of the arrow shaft, the size, shape and material of the arrowhead and the fletchers, even the shooting style, all contributed to how far a bow could fire an arrow.

There are numerous references in the literature to 'bow shot' as a term to indicate distance.[78] From this, it is clear that there was a general understanding among the ancients of how far a bow could shoot. Unfortunately, there is little information to help us understand the ancients' 'bow-shot' in modern measurable terms. The only indication of bow range in antiquity is from an inscription from Olbia, an ancient Greek city in modern Ukraine. The inscription commemorates a certain Anaxagoras' victory in the flight shooting (long distance shooting) competition *circa* 300 BC.[79] The victorious archer managed a staggering distance of 515 metres. The bow used was undoubtedly a Scythian bow, based on the geographical location of Olbia.[80] This has to be viewed as an extreme range of a period bow and not the norm. Even under the Ottomans in the seventeenth and eighteenth centuries, where archery, flight shooting and arguably the bow itself reached their zenith, 500 metres and above was considered exceptional enough to have one's name engraved into stone or marble columns to commemorate the skill and achievement. These long-distance shots were achieved with specialized arrows that were tapered and very short to reduce friction and make them as aerodynamic as possible. They were thus not suitable for warfare or accuracy. We may assume that the shot made by Anaxagoras at Olbia was done with similar arrows to achieve the recorded distance.

From this we can deduce that the bow's effective military range had to be significantly less than 500 metres. There is one more reference that might give some sense of the range of ancient bows, Strabo mentioning that in the first century BC, Mithridates VI of Pontus shot an arrow a little further than a stadium (176 metres for an Olympic stadium, 185 metres for Attic).[81] This distance is very similar to the range given by the fourth-century AD military writer Vegetius, who advocates that Roman auxiliary archers should practise shooting at targets placed 177 metres (600 Roman feet, 580 modern feet) away.[82] It is thus reasonable to assume that the bow had a range of approximately 150–200 metres. In order to achieve such a distance, the arrow had to be shot at an elevated angle, or 'underhand', as it is known in the sphere of archery. These arching or parabolic shots sacrificed accuracy and the force of impact of the arrow for distance.

When archers shot their arrows in a straight or horizontal trajectory, they were much more accurate and arrows struck their targets with a greater amount of force. The range, however, would be much shorter since the gravitational force upon the arrow had a greater effect. I am of the opinion that Greek and Roman archers from the period under discussion focused more on sniping at individual targets by shooting at a much closer range, at a horizon trajectory, as opposed to long-range parabolic shots (see 'Fighting at Close Range' in Chapter 2.2, Archers in Combat). From personal experience as an historical archer (as well as the experience of other archers), and the assessment of literary, iconographical and archaeological evidence, these horizontal trajectory shots were effective anywhere from a very close range of 10 metres to a maximum of 70 metres.

Drawing the Bow

In modern archery, at least in the West, most archers make use of the 'Mediterranean' draw. This method of drawing the bow has the archer grip the bow with the left hand and draw the string with the right hand (assuming the archer is right-hand dominant). The string of the bow is drawn with the index, middle and ring finger. The arrow is placed on the left side of the bow, while the nock of the arrow (the back of the arrow shaft which is placed onto the string) is held in place between the index and middle finger of the right hand (**see plate 8**). The drawing hand is thus held vertically, with the index finger on top of the hand and the little finger at the bottom. This is also how our modern culture often depicts Greek and Roman archers in films, book illustrations and other artwork.

While the Greeks were certainly familiar with the Mediterranean draw, as the Alexander Sarcophagus and the Nereid monument demonstrate, a vast number of depictions show that archers in the Graeco-Roman world shot their bows quite differently.[83] The arrow is often placed on the right-hand side of the bow and the drawing hand (right hand) held horizontally with the thumb facing inward (to the left) and the little finger facing outward (to the right). The string is drawn with the thumb that is hooked around the string and the index finger (and or the middle finger in other variations of the draw) wrapped around the thumb for support (**see plate 9**). This 'thumb draw' is very familiar to those who practice traditional 'Asiatic' and horseback archery. This type of draw was used in some shape or form by the Egyptians, Assyrians, Scythians, Romans, Byzantines, Sassanids, Huns, Avars, Mongols, Turks, Chinese and Koreans. This form of shooting was certainly known to the Greeks, based on representations in art. A prime example is once again the fifth-century amphora of Odysseus slaying Penelope's suitors.[84] Odysseus is portrayed drawing the bow with the arrow on the right side of the bow and what appears as a thumb draw technique pulling back the string (**see plate 10**). Another example would be the sculpture of Paris on the Aphaia West Pediment (500–490 BC). While the arrow and bow are missing, the drawing hand is held in a horizontal position and the draw has to be interpreted as a type of thumb draw. Paris is again portrayed drawing the bow with a thumb draw by the Niobid Painter in the 'Death of Achilles' vase scene (460 BC). Another thumb draw technique can be seen on a *kantharos* which shows Heracles fighting Amazons. To the right of Heracles is a kneeling Amazon drawing her bow, aimed at the Greek hero, with a clearly depicted thumb draw.[85] The Berlin Painter also depicts Heracles fighting Amazons, where a female archer draws a bow with her hand held horizontally, which can certainly be interpreted as a type of thumb draw.[86] There is also a fifth-century kylix drinking cup showing a 'Scythian' archer shooting over the body of a fallen hoplite, where the draw technique can once again be seen as a type of thumb draw.[87]

Archaeological evidence possibly also points to the Greeks making use of the thumb draw. As we have seen, ancient bows were very powerful. In order to draw the bow with the thumb, a leather thumb pad or a thumb ring of bone, wood or metal was often used to protect the thumb. Just such a thumb ring was discovered at the Mycenae citadel, dating from the late Bronze Age.[88] While not conclusive, since the ring could be interpreted as a gift from a foreign ruler, it does establish the link between ancient Greece and the thumb draw as far back as Mycenaean times.

It is clear that archery in the Graeco-Roman world was influenced by the Scythians and other nomadic tribes roaming the steppes. It should thus not be surprising that the Greeks adopted the thumb draw as their method of drawing the bow. The thumb draw was particularly suited for chariot and horse archers, the Scythians being the latter. With the Mediterranean draw, the arrow is placed on the left side of the bow. The arrow is simply resting between the upper knuckle of the index finger and the bow itself. If the archer was in a fast-moving chariot or upon a galloping horse, the arrow would bounce around on the shaft of the bow as nothing kept the arrow in place. The thumb draw, however, solves this problem by placing the arrow on the right-hand side of the bow. The arrow is kept on the string by the index finger and thumb of the drawing hand. At the same time, the index finger holds the arrow in place on the bow by pushing the arrow onto the bow shaft. No matter the amount of movement, the arrow remains in place, making it ideal for archers shooting from a fast-moving object such as a chariot or horse. It is thus understandable why this type of draw was used by cultures whose martial arts were centred on horse archers.

The reason that the arrow is placed on the left side of the bow when the Mediterranean draw is used is that when the three fingers pull the string, they rotate the string clockwise, which in turn helps to rotate the arrow towards the bow and helps to keep the arrow in place. While not impossible, it is more difficult to keep the arrow in place when drawing the bow with the Mediterranean draw if the arrow is on the right side of the bow, especially with heavy military-weight bows (75lb and upward). When the bow is drawn with the thumb, it rotates the string anti-clockwise and thus pushes the arrow (which is placed on the right of the shaft) towards the bow. It is thus also a fairly good indicator that a type of thumb draw is being used if the arrow in the iconography is placed on the right side of the bow.[89]

There were probably several different draw techniques, but it does appear that, contrary to popular belief, archers in the Graeco-Roman world were very familiar with and did use the thumb draw technique.

2.1.2 Quivers and the Arrows

Quivers

Archers in the Graeco-Roman world used a variety of quivers (φαρέτρα). Again, we have to rely on iconography. Several images show the archers with cylindrical back quivers. These back quivers appear most often in a Greek or Persian context. Interestingly, several of these have quiver caps (*pôma*/πῶμα), which would both protect the arrows and stop them from falling out when the archer was running or moved through difficult terrain. The late seventh-century Skyphos crater bowl and a sixth-century amphora from the Metropolitan Museum of Art serve as examples. In both images, Heracles has a quiver on his back, with the quiver cap suspended from it by a thong.[90] It is also attested in the *Iliad*:

> 'Then opened he [Pandarus] the lid (*pôma*) of his quiver, and took forth an arrow, a feathered arrow that had never been shot, freighted with dark pains; and forthwith he fitted the bitter arrow to the string.'[91]

There are also plenty of images of side quivers, which either hung from the bowman's belt or from a baldric draped over the shoulder. Scythians, Persians, Greeks and numerous characters from Greek mythology are often portrayed with a type of side quiver. In the majority of the images, these quivers are suspended on the left side of the archer. The *gorytos*, the combined quiver and bow case, was popular among both Persians and Scythians, but less so among the Greeks. An archaeological find of a golden *gorytos* from the excavated Macedonian Vergina Tombs serves as a magnificent example. The *gorytos* dates from the latter half of the fourth century BC and in all likelihood belonged to one of the wives of Philip II of Macedonia. While the *gorytos* was not very popular with the Greeks, that is probably because the large quiver and bow case were too cumbersome for the Greek archer's fighting style (see Chapter 6.3: Cretans at War). There are several images of Greeks tying their bows to their quivers when not in use.[92]

Arrows

Arrows came in all shapes and sizes (**see plate 11**), with oth bronze and iron arrowheads in use. Some arrowheads were socketed, while others were tanged. Arrowheads were often of the trilobate type, but some were double-bladed broadheads with large extended barbs. Different arrowheads served different purposes. Smaller, narrower heads could better penetrate armour, while larger barbed heads inflicted more severe wounds and caused greater tissue damage. The trilobate head, which dates as far back as the Bronze Age, inflicted a 'star wound' which was a nightmare for a surgeon to treat as it was difficult to stop the bleeding and stitch up the entry point. Some were a narrow leaf-shape bilobate-type and at times had a single hook-like barb protruding a little behind the arrowhead. Barbs made it difficult to remove the arrow, and even a simple flesh wound would have need of surgery, as the following quotation from the medical treatise *De Medicina* shows:

> 'Nothing penetrates so easy into the body as an arrow, and it also becomes very deeply fixed. And this happens both because it is propelled with great force and because it is sharply pointed. Hence it is more often to be extracted through a counter opening than through the wound of entry, and especially so because it is generally furnished with barbs which lacerate more when drawn backwards than if pushed through a counter opening.'[93]

The devastating effects of arrows are once again illustrated in Plutrach's account of Carrhae:

> '[The Romans] died no easy nor even speedy death. For, in the agonies of convulsive pain, and writhing about the arrows, they would break them off in their wounds, and then in trying to pull out by force the barbed heads which had pierced their veins and sinews, they tore and disfigured themselves the more.'[94]

Nine arrow wounds appear in the pages of the *Iliad*, and in every case the wounded man could not remain on the battlefield but had to retreat from the action.[95] The following verse from the *Iliad* highlights both the value of battlefield surgeons and the prevalence of arrow wounds during the period: 'For a physician is a man worth many others, for cutting out arrows and applying soothing remedies.'[96]

Arrowheads could also be covered in poison (I have already mentioned the link between our modern term 'toxin' and the Greek word for bow). Scythians were apparently infamous for using poison on their arrows:

> 'They say that they make the Scythian poison with which they smear arrows, out of the snake ... with the juice of the snake, and so make a deadly poison.'[97]

Poison arrows, possibly the earliest form of biological warfare, had the potential to make the most superficial of wounds fatal. While poison arrows held certain physical and psychological advantages, they also held a nefarious reputation of treachery and dishonour; it appears that poisoned weapons in war were not in common use among the Greeks and Romans. This reputation seems to date back as far as Homer. In the *Odyssey*, Odysseus set out to find the 'man-killing drug' for his arrows, but Ilus, the leader of Ephyre, where the hero went in search of the poison, refuses to give it to him because he feared the wrath of the gods.[98]

Arrow shafts were of both wood and reed. The rattan reed, for instance, is mentioned in the first century AD as being used in the Roman and Greek world for arrow shafts, and the *Iliad* also makes mention of reed shafts.[99] Lighter reed arrows could be shot greater distances, but heavier wooden arrows, albeit having a shorter range, had more penetration power.

2.1.3 Defensive Equipment

Body Armour

As mentioned, archers, ike other light troops, were very lightly armoured if at all. Most iconographical representations of archers depict bowmen as wearing no armour. There are, however, a few images of archers clad in light armour. Some images show archers wearing the *linothōrax* (λινοθώραξ), which was basically a tube constructed of various layers of linen that fitted around the wearer's waist, with shoulder flaps or *epōmides* (ἐπωμίδες) (**see plate 12**). These straps were attached to the back of the tube, which folded over the shoulders and were fastened in front of the chest. Several layers of cut strips of overlapping linen, known as *pteruges* (πτέρυγες), were attached to the bottom of the tube, which gave protection to the thighs and groin without restricting movement. The material used for its construction is uncertain. Scholars have proposed a variety of theories, such as hide covered with linen, metal covered with linen or simply several layers of linen.[100] I do not believe it to have been constructed of hide or metal and simply covered in linen. The name itself implies linen to be the main material for the construction of the armour. The experimental archaeology study *Reconstructing Ancient Linen Body Armor: Unraveling the Linothōrax Mystery* has shown that a *thōrax* constructed of solely linen makes for very effective armour and could protect its wearer against attacks from a variety of weapons.[101] The archer, interpreted as Heracles, on the East Pediment of the Temple of Aphaia is wearing *linothōrax*. The Nereid Frieze displays two kneeling archers shooting up at besieged city defenders. The bowman on the left is wearing a *linothōrax*, while the archer on the right is completely unarmoured. There are also several Persian and Amazonian archers in Greek art wearing this type of armour.[102]

This *linothōrax* was well suited for light infantry, as it was flexible and light. Archers could move freely over obstacles and difficult terrain, this type of armour posing very little restriction to body movement. I found that while I wore a reconstructed version of the *linothōrax*, my archery was not affected at all and

I had a full range of movement, being able to jump, roll and climb due to the lightness and flexibility of the armour. As we have seen, archers – whether they were mercenaries or citizens – often came from the poorer strata of society and did not have deep pockets to invest in expensive hoplite armour. The *linothōrax*, however, was relatively cheap and far easier to construct in comparison to metal armour, and clearly some archers were able to afford this type of armour. It was also very popular with the heavy infantry, there being far more representations of hoplites than archers wearing this type of protection.

Helmets

Some depictions also show archers wearing helmets. The Nereid Frieze depicts a Greek archer with a crested Attic helm.[103] The Pithos Painter's depiction of a kneeling archer is also wearing a crested Attic helmet, but is otherwise completely naked.[104] There is also the depiction of a kneeling archer wearing a *konos* or *pilos* helmet.[105] Both these types of helms suit archery as they are open-faced and do not restrict the archer's vision. There is also the depiction of a very well equipped bowman on a kylix from the Etruscan city of Vulci, attributed to the painter Oltos, wearing a closed-faced Corinthian helm.[106] It is interesting to note that the archer has tilted back the helm so his vision is not restricted while shooting his bow.

Shields

Another form of defensive equipment that could be used by archers was the shield. There are a few rare depictions of bowmen carrying shields. The kneeling archer by the Pithos Painter discussed above has a shield propped up against his leg while shooting his bow. The helmed archer in the Nereid monument also carries a shield. The image is peculiar as the shield is a large *aspis* type, typically associated with hoplites. It would not be possible to draw a bow while holding such a large shield. This is indeed the case with this archer, as the shield is not carried on the arm but slung over his back. A similar image on a bell krater by the Altamura Painter shows two hoplites fully clad in armour, each carrying a large *aspis* shield.[107] The hoplite on the left, however, has a quiver hanging from his left hip. One might assume that these types of archers were multi-purpose soldiers who could rain arrows upon the enemy from a distance and then engage in a melee, perhaps even joining the ranks of the phalanx as these shields and other equipment would allow them to do so. Another possibility might be that they were *toxarchoi*, officers of an archer corps, being better paid and thus able to afford better equipment. It is interesting to note that the Athenian professional archer corps could earn enough to buy and keep their own slaves, as attested in an Athenian naval catalogue where two archers are recorded as boarding with a slave each.[108]

Smaller shields could, however, be used in conjunction with the bow. There are a few rare images of archers shooting their bows while carrying shields attached to their forearms. A red-figure kylix, now in Basel, has a naked archer with greaves and a Scythian-styled felt hat, drawing his bow with a small shield attached to his left forearm.[109] A very similar figure appears on another kylix attributed to the artist Epiktetos;[110] the naked archer draws his bow whilst kneeling and has a small shield attached to his left forearm. Finally, the image of the well-equipped archer on a kylix from Vulci by the painter Oltos, mentioned above, serves as a fine example of all the defensive and offensive equipment discussed. As well as a *linothōrax*, greaves and a Corinthian helmet tilted back on his head, the bowman has a shield attached to his left forearm, with the top of a bow seen sticking out from behind the shield and a side quiver hanging from his left hip.[111]

Archers carrying shields are also attested in literature, generally being associated with Cretans (see Chapter 6.2.2: Shields). Both Xenophon and Polybius mentioned Cretans carrying small shields.[112] Xenophon calls these shields *peltai* (πέλται) (**see plate 13**).

2.2 Archers in Combat

Fighting at Close Range

The idea of archers on the battlefield, very much thanks to Hollywood, conjures images of massed bowmen releasing several volleys of arrows from a long distance away, shooting in parabolic arcs so that the missiles fall upon the enemy from above. Images of the English longbowmen and their victories over the French at Crécy (1346) and Agincourt (1415) come to mind, or the Persians, whose arrows at Thermopylae (480 BC) were so numerous they 'blot out the sun'.[113] While this type of long-distance shooting of massed arrows in an arc did exist in the ancient world, particularly among the Achaemenids and later Imperial Romans, archers serving in Greek and Republican Roman armies appear to have fought in a very different manner. Depictions of archers during our period are never shown shooting arrows upwards in a parabolic arc to gain distance. They rather shoot their bows at a straight or horizontal trajectory, directly at their targets. On the battlefield, they thus acted as snipers, picking particular targets at which to shoot their arrows, rather than shooting them in a general direction at a clump of men. Archers are also continuously portrayed in the thick of battle, stationed right behind hoplites or alongside them, at times taking shelter behind the shields of their heavy infantry comrades. While this could be interpreted as simply due to the restrictions of the painter's work surface or artistic interpretation, the literary evidence supports the fact that archers fought at close range, picking off individual opponents, as opposed to long-range shots at massed groups of enemies.

Going back to the pages of the *Iliad*, archers are almost always described as sniping at opponents picked out from among the fighters on the battlefield. Shots are often taken when their opponents are in the thick of battle and engaged in melee fighting. It would be impossible to identify a particular person 150–200m away among a mass of enemies. Shooting in a parabolic arc would reduce the accuracy of the shot and the archer would risk hitting his ally and friendly troops. Parabolic shots also take several seconds to hit the target due to the long arch the arrow travels. In the chaos of melee combat, it is highly unlikely that the intended target would be in exactly the same position when the arrow was shot, increasing the chances of missing one's intended target and at the same time exposing one's allies to 'friendly fire'. Furthermore, archers in the *Iliad* are described as shooting at such a close range that the bowman and his wounded victim are often in verbal communication with one another. In order for these men to exchange words through the din of battle, they surely had to be a maximum of 50 metres apart, and in all likelihood much closer. Looking at archery from a more practical point of view, a bow's hunting range (which according to the ancients served as good practice for war) is generally between 15 and 40 metres, with accuracy beyond this range diminishing tremendously.[114]

The most telling evidence comes from Onasander in his *Strategikon*, who clearly advocates that archers should shoot the arrows in a straight trajectory as opposed to parabolic shots:

> 'If the archers are positioned in front of the army they shoot their arrows straight at the bodies of the enemy just like at a target. But if they are positioned behind the other ranks or in the middle, they will shoot upwards, so that the arrows have impetus only for their flight upwards, and afterwards have spent their force so that even if they fall directly in the heads of the enemy they cause little damage.'[115]

From the above, it is clear that when shooting in a straight trajectory over a shorter distance, arrows struck the intended target with greater kinetic energy and penetrative power.

The strategic deployment of archers alongside other troops also supports the idea of close-range sniping as opposed to mass volleys from long distance. As previously mentioned, Herodotus' account of the Battle of Platea has the Athenian archers bring down the horse of Persian cavalry commander Masistius.

These archers had to be shooting at a close range, as we are told that 'when he fell, [he] was straightaway set upon by the Athenians [hoplites]. His horse they took then and there, and he himself was killed fighting.'[116] Masistius must have been unhorsed very close to the Athenian line for the hoplites to rush forward and spear him. The archers, being stationed alongside the hoplites, had to be shooting in a straight trajectory if Masistius' horse was brought down so close to the Athenian line, as parabolic or arching shots are only intended for long-distance shooting.

Archers are often paired with other melee or close-range troops such as javelineers. Alexander the Great, for instance, almost always paired his archers with the Agrianes, a Thraco-Macedonian people who served as an elite light infantry armed primarily with javelins.[117] Javelin throwers had a much shorter range than archers and had to cast their missiles at a very close proximity to the enemy, somewhere between 10 and 20 metres. If archers were shooting from a long distance away in a parabolic arc, they would risk hitting their own javelinists. Archers were also at times paired with offensive melee cavalry. Again, archers would risk hitting their own troops if they shot parabolic shots while their own horsemen were engaged in hand-to-hand combat with the enemy. Furthermore, had these archers been deployed at long range and shooting parabolic shots, the other troops they supported would be between them and the enemy; their view of the enemy would thus be obscured, making it difficult for them to place their shots. Sniping at individual targets at close range, scattered among the troops they were supporting, would negate the above-mentioned problems.

When archers came up against heavy infantry, they had to contend with opponents who were generally well armoured and carrying large shields. Heavy infantry armour was very effective against arrows, and Davis highlights that mass volleys would be wasteful upon period armour, the Battle of Marathon (490 BC) serving as a prime example.[118] Consequently, archers thus had to pick their shots, aiming carefully at exposed body parts. They thus had to be in a close proximity. Arrows did not need to kill to neutralize a soldier on the battlefield, and since the body and head were often clad in armour, exposed arms and legs became the most likely targets. The arrow that hit Diomedes in the foot might thus not have been due to poor aim but instead well-placed and intended. Paris took out another Achaean, Eurypylus, by shooting him in the leg:

> 'He [Paris] drew his bow against Eurypylus, and smote him with an arrow on the right thigh; and the reed of the arrow brake, yet was his thigh made heavy.'[119]

The Peribolos Tomb, which held the bodies of Spartan soldiers who died fighting Thrasybulus' troops in his rebellion against the Thirty Tyrants (403 BC), contained a skeleton of particular interest. Lodged in the right leg of the skeleton were two bronze arrowheads.[120] The fact that there were two arrowheads in the leg supports the idea that legs were targets for archers and other light troops. Xenophon provides us with an account of Thrasybulus' light troops, including the archers, inflicting heavy casualties upon the Spartan hoplites:

> 'And the light troops, rushing forth at once, set to throwing javelins, hurling stones, shooting arrows, and discharging slings; then the Lacedaemonians, since many of them were being wounded and they were hard pressed, gave ground, though still facing the enemy; and at this the latter attacked much more vigorously.'[121]

Further evidence that the legs were prominent targets for archers was a shield attachment called an arrow-curtain, which was made of leather or linen. As the name implies, it was a rectangular piece of material attached to the apsis shield, hanging like a curtain from the lower end to provide protection to the legs against missiles. There are several examples of these in Greek art; on the Nereid monument, the archer bearing a shield upon his back has an arrow-curtain attached to it.[122] In a Roman context, the Republican *scutum* shield was oblong and gave ample protection to the legs, and therefore never required such a device. The Romans at the Battle of Carrhae (53 BC) were initially content to wait out the arrow storm of the Parthians behind their *scutum* shields until the enemy's arrows ran out. This, however, would not happen, as Surena, the Parthian general, had brought with him a caravan of 1,000 camels carrying vitually limitless numbers of arrows:

> 'Now as long as they had hopes that the enemy would exhaust their missiles and desist from battle or fight at close quarters, the Romans held out; but when they perceived that many camels laden with arrows were at hand, from which the Parthians who first encircled them took a fresh supply, then Crassus, seeing no end to this, began to lose heart.'[123]

As with exposed legs, the upper extremities were prime targets, with five out of the nine non-fatal arrow wounds mentioned in the *Iliad* being in the shoulder or arm.[124] The well-defended heavy infantry thus also dictated the need for period archers to shoot at close range.

Davis highlights that the tactic of mass arrow volleys was only effective if military leaders were able to deploy large numbers of archers, such as the Near

Eastern empires of Persia and Parthia evidently could.[125] Classical Greece never had the numbers to field massed archer formations which could 'blot out the sun'. The largest expeditionary force of the Peloponnesian War, for instance, was that which Athens sent to Sicily, which consisted of 134 triremes and more than 60,000 troops, of which only 480 were archers.[126] As mentioned, Roman Republican armies predominantly relied on mercenaries, often supplied by allied states.[127] Hiero II of Syracuse, for example, sent 1,000 slingers and archers to the Romans, while Eumenes of Pergamon provided a force of 500 Cretan archers – again, hardly enough to 'blot out the sun'.[128] It was only after Alexander brought the Achaemenid Empire to its knees and the Successor States gained access to large pools of archers that the Hellenistic Successors fielded archers in their thousands.

Naval Warfare

Archery formed a prominent component of naval warfare. According to the Decree of Themistocles (lines 23–26), the ideal number of fighting personnel on each Athenian trireme was ten hoplites and four archers. The complement of four archers seems to have remained the standard throughout the Peloponnesian War, as is evident from Thucydides.[129] This was, however, not set in stone, an inscription dating from the same period stating that a particular thirty ships were each provided with forty hoplites, ten archers, ten *peltasts* and an additional five volunteer *epibatai* (marines, generally heavy infantry).[130] These inflated numbers most likely suggest an amphibious assault as opposed to an open sea battle, as highlighted by Thucydides' account of the sea battle in the Syracusan harbour:

> 'Many archers and javelineers will be on deck and a mass of hoplites, which we would not employ if we were fighting a battle in the open sea, because they would hinder us through the weight of the ships in exercising our skill.'[131]

On the decks of a ship, archers had great offensive capability. Greek naval warfare was predominantly based on outmanoeuvring the enemy and ramming their ships. Hoplites were thus reserved for boarding actions, which were less frequent. Archers, on the other hand, could snipe at the enemy from the deck of a ship while the ships performed their tactical manoeuvres, a prime target being the enemy rowers. Targeting the oarsmen had the obvious advantage of disrupting the rhythm and synchronicity of the oar strokes if they were hit. This in turn would compromise the speed and dexterity of the ship, making it vulnerable to being rammed. The threat posed by archers to the rowers was quickly recognized, and leather curtains, not unlike those sometimes attached to shields, were erected on the sides of warships to protect the oarsmen from missiles.

Archers also came into play after a successful ramming action when the rammed ship sank, the crew who tried to swim to safety finding themselves unfortunate targets; according to Diodorus, 'these would then be shot by archers'.[132]

Even long after the Athenian golden age of the fifth century BC, the city-state's naval archers were still able to provide a good account of themselves against the Celtic invasions of the early third century BC. According to Pausanias, at the Battle of Thermopylae (279 BC) the Athenian navy provided support to the Greek coalition against the Celtic forces of Brennus by sailing their ships as close to the shore as possible and unleashing missiles into the flanks of the 'barbarians':

> '[T]he Athenians on the triremes, with difficulty and with danger, nevertheless coasted along through the mud that extends far out to sea, brought their ships as close to the barbarians as possible, and raked them with arrows and every other kind of missile.'[133]

During boarding actions, archers could play both a defensive and an offensive role. In the event of boarding an enemy vessel, archers could lay down suppressing fire upon the enemy while the hoplites made their crossing to the opposition ship. In the case of being boarded, archers could shoot their arrows at both the enemy hoplites, hampering their boarding action, and enemy missile troops.

Having archers on the deck of a warship was clearly advantageous, and a captain would take to the seas without them at his own peril.

Siege Warfare

Archers played an important role during sieges, both in the defence of a city and in attacking it. A city's greatest defences were its fortified walls and towers. Those who ventured out to attack them faced a myriad of missiles (especially arrows) being shot from the battlements. The obvious main task of a city's defenders was to keep the enemy from reaching the walls. Archers, with their range advantage, thus played a vital part in a city's defence. As on the battlefield, archers served predominantly as snipers, as opposed to shooting mass volleys of arrows. Perched up high in the towers or mounted on the curtain wall, archers held an obvious height advantage when shooting down at the enemy. From up high they could far more easily pick out their targets. Along with the height advantage, they were also better defended in the towers and behind the parapets of the curtain wall. Arrow slits in walls and towers were typically just 10–15cm wide on the outside but 50–60cm wide on the inside. This gave the defending archers the room to comfortably draw their bows and loose their arrows through the slit, but made it far more difficult for the enemy to send their own missiles through these slits at defending archers. It is interesting to note that these arrow slits were particularly designed for archers and were not suited for slingers or javelinmen, highlighting the importance of archers in the defence of a city.

The towers along the curtain wall were one of the greatest defences of a fortified city. Vitruvius, the Roman architect and engineer who wrote his celebrated treatise *On Architecture* during the first century BC, states that 'the distance between each tower should not exceed an arrow's flight; so that if, at any point between them, an attack be made, the besiegers may be repulsed'.[134] Based on fortifications of this period, towers were typically 50–60 metres apart so that the archers within could effectively cover the intervals.[135] This supports the idea that the effective bow range was approximately 50 metres.

The need for archers in the defence of a city during a siege was highlighted by the Milesians in their revolt against Athens. Expecting their previous overlords to send forces to besiege their city, the Milesians sent gold to hire archers (possibly Scythians) from Pontus.[136]

For the besiegers, who had to reach the walls to either scale them with ladders or breach them with rams, archers posed a serious threat. Fifty metres was a long distance to cover when pushing a ram or carrying siege ladders under a salvo of missiles. When approaching a wall between two towers, these soldiers would be exposed to missiles from the front as well as the flanks. Missiles would also come from various levels and angles, as walls and towers had several levels from which arrows could be shot. An attacker would need both hands to push or carry siege equipment, meaning they would not be able to carry a shield, making them far more vulnerable to the defenders' missiles. They would also be prime targets for missile forces because they posed the greatest threat to the city. Even a soldier who carried a shield would be unable to protect himself sufficiently from arrows shot from all sides. Military engineers and sappers had to build and construct their own defences, and these labour-intensive jobs of building and digging were often done when wearing no armour, making them vulnerable to the defenders' arrows and other missiles. It is thus understandable why sieges could last months and why starving out the city or going through the labour-intensive task of undermining the walls were preferred to storming the city.

For the besiegers to breach or mount the walls they needed a means to counter the defenders, whoch was done through a salvo of their own archers and other missile troops. Archers, along with other light troops, played an important role in Alexander's siege of Gaza:

> 'He [Alexander] at once ordered his men to attack the wall, which was made of earth and of no great height, and to place their ladders against it on all sides; while his slingers and archers and javelin-men, at the moment of the infantry attack, showered volleys upon the first line of defence on the walls, and missiles were hurled from the catapults, so that the wall was soon cleared of defenders by the great shower of missiles, and thus the setting up of the ladders and the ascent of the Macedonians to the wall were rapidly accomplished.'[137]

During the siege of Syracuse in 213–212 BC, Polybius and Livy state that the Romans relied heavily upon archers and other missile troops to clear the battlements of the city.[138]

In an attempt to make the playing field more equal, besiegers erected mobile siege-towers, which often towered (no pun intended) over the defenders' walls. The towers used by Alexander in his siege of Tyre (332 BC), for instance, were reported to be 45 metres high. These towers, which could have portholes and several storeys of their own, served as effective platforms from which besieging missile troops, particularly archers, could shoot their missiles to clear the walls so the besiegers could approach them. Diodorus' account of the Carthaginian siege of Selinus (409 BC) serves as a prime example of the effectiveness of siege towers as a platform for archers and other missile troops:

> 'He [Hannibal] set up six towers of remarkable height and brought forward the same number of iron-shod battering rams against the walls; in addition to these, he used great numbers of archers and slingers to drive back the combatants on the battlements.'[139]

Archers could also shoot incendiary or flaming arrows. These were used by both sides; the besieged attempted to set alight the wooden siege towers and other siege equipment, while the besiegers attempted to burn down wooden ramparts and cause confusion by setting fire to buildings inside the city. Herodotus' account of the Persian siege of the Athenian Acropolis (480 BC) has Persian bowmen successfully using incendiary arrows against the wooden barricades erected by the defenders:

> '[T]hey wrapped arrows in tar and set them on fire, and then shot them at the barricade. Still the besieged Athenians defended themselves, although they had come to the utmost danger and their barricade had failed them.'[140]

The importance of archery in siege warfare can further be seen in the preparations of the city of Sinope in anticipation of its siege by Mithradates II of Pontus. The city of Rhodes provided aid to Sinope in the form of 100 talents worth of bowstrings to help with the siege preparations against Mithradates.[141]

Sieges were a common feature throughout the ancient world, with archery an essential element of siege warfare for both the defenders and the attackers.

A Counter to Light Troops

During battles, light troops often ended up fighting the enemy's light troops. Battles and skirmishes in difficult terrain, where the heavy infantry could not deploy, were fought by the light troops, which decided the outcomes of such clashes.

As already mentioned, pitched battles were decided by the heavy infantry, but the fighting commenced with a clash of the light troops of the two opposing armies. Thucydides makes it seem like the actions of these light troops were fairly inconsequential: 'And first the stone throwers, slingers and archers fought each other in front of the main formations. Each side routed the other in turns as is normal in combat between light-armed troops.'[142] Nevertheless, it should be stressed that light troops could and on occasions did defeat heavy infantry. Light troops could easily outflank the enemy heavy infantry and attack their vulnerable flanks and rear while they were otherwise engaged in melee. However, light troops very much served as defensive shields for the heavy troops in these set battles, so it was essential for each side to have light troops to counter those of the enemy.

When light troops opposed each other, archers held an advantage over javelinmen, who were the predominant light infantry of the Graeco-Roman world. Firstly, the bow outranged the javelin. As with archers, javelineers threw their missiles in a straight trajectory, sniping at individual targets, so their effective range was only 10–20 metres. Archers thus had a range more than double that of the javelin throwers. Arrows were far smaller than javelins and when shot from a bow travelled at a much greater velocity, making them more difficult to dodge or block with a shield. Archers could also carry a far greater amount of ammunition than a javelinist could. While there is little indication of how many arrows were carried by a typical archer from the period, English medieval archers carried two sheaves of arrows which totalled twenty-four (a sheaf being twelve arrows).[143] Byzantine quivers during the tenth century, however, carried up to sixty arrows.[144] Again, there is very little reference to how many weapons a javelineer carried. However, the Roman javelinists, according to Livy, carried seven javelins onto the battlefield.[145] There is a big difference between seven javelins and sixty arrows, meaning that archers had much greater staying power in range fighting than javelinists. Thucydides' account of Demosthenes's invasion of Aetolia (462 BC) has the bowmen serving as a 'shield' for the hoplites and giving a good account of themselves against the Aetolian javelinists:

> 'But as long as the Athenian archers had arrows and were able to use them, they held out since the Aetolians were light-armed and retreated when they were shot at.'[146]

However, when the *toxarchos* leading them was slain, the archers scattered and the hoplites no longer had light troop protection from the Aetolian javelinmen. As a result, Athens suffered one of its worst defeats of the Peloponnesian War.

A Counter to Cavalry

Archers served as an effective anti-cavalry unit. As far back as the *Iliad*, we see Nestor being thrown from his chariot after Paris brings down one of his horses, having 'smitten him with an arrow upon the crown of the head ... the deadliest spot. So, stung with agony the horse leapt on high as the arrow sank into his brain.'[147] These were by no means weak or lightweight bows if an arrow could penetrate a horse's skull, even shot at close range.

The archers at the Battle of Plataea have already been mentioned, those of both sides serving as a counter to cavalry. It was the Athenian archers who brought down Masistius' horse, leading to the death of the Persian officer.[148] The second time archers were mentioned in the battle was when the Spartan general Pausanias, being hard pressed by the Persian cavalry, requested Greek archers to be sent to the Spartans' aid.[149] The effectiveness of Greek archers against cavalry at Plataea is further enhanced by Simonides of Ceos' epitaph:

> 'This bow, resting from tearful war, hangs here under the roof of Athena's temple. Often mid the roar of battle, in the struggle of men, was it washed in the blood of Persian horsemen.'[150]

Archery in the Hellenistic era was dominated by Cretans, who gave an excellent account of themselves as anti-cavalry troops, which is discussed in greater detail in the chapter dedicated to light troops from Crete (see Chapter 6: Cretan Archers).

Support Units and Combined Arms Combat

Archers often fought as support units in combined arms tactics. In the *Iliad*, Teucer often pairs himself with his heavy infantry brother, Ajax, kneeling beside him and shooting arrows at the enemy from behind the shelter of his brother's huge shield:

> 'And Teucer ... tensing his reflex bow and lurking under the wall of giant Ajax's shield. As Ajax raised the rim, the archer would mark a target, shoot through the lines – the man he hit dropped dead on the spot – and quick as a youngster ducking under his mother's skirts he'd duck under Ajax's shield and the gleaming shield would hide him head to toe.'[151]

This archer–hoplite pairing seems to have been a common practice in early hoplite warfare, with numerous iconographical evidence showing archers throughout the archaic and early Classical period taking cover behind the shields of the heavy infantry while shooting their bows.

Throughout the Classical period, archers served as a support unit for hoplites. I have already mentioned the archers fighting alongside the hoplites at Platea and those who protected the Athenian hoplites against the Aetolian light infantry in 462 BC. Demosthenes specifically grouped his Athenian hoplites with archers and successfully defended the Pylos beach (425 BC) against an amphibious assault from a numerically superior Spartan force.[152] Due to the difficult terrain, the small number of hoplites (sixty in total) and the restrictions of hoplite warfare, we may assume that it was the archers who played the most significant role in repelling the Spartan forces.

Archers often served as support troops for Alexander the Great's assault infantry. In Arrian's *Anabasis*, Alexander almost always grouped his elite Agrarian javelinists with his archers. Arrian mentions this pairing fifteen times, while each of these troops only appear once on their own. Alexander's Successors also paired their Cretan archers with light assault troops. Both Philip V of Macedon and Antiochus the Great paired their Cretans with Trallians, who were 'swift-footed Illyrians' who 'would be employed in sudden onsets and rushes'.[153]

2.3 Conclusion on Archers

There were numerous other tactical roles filled by archers, such as operations in difficult and uneven terrain, ambuscades, forced marches and scouting tasks, which were typically shared by all light troops and will be discussed in more detail in a separate section below.

Certain prejudices against archers existed in the ancient world, and the idea that archery played only a minor role in the Graeco-Roman world persists even today. However, taking a more in-depth look at the sources, it becomes clear that archery was an integral part of Greek and Roman warfare. While archery was not the dominant arm of Greek and Roman armies, the archer and his bow played a vital part on ancient battlefields. Only a foolish warrior would brush off the deadly arrow as inconsequential, and a general fielding an army without archers would surely be poorer for it.

Chapter 3

Stone Throwers: Sphendónitai/ Funditores and Lithoboloi

Slingers

The sling (*sphendónē* in Greek/*funda* in Latin) was used as a weapon of war from as early as 10,000 BC, and its use continued deep into the Middle Ages.[1] Even after the arrival of gunpowder, the sling remained in use up to the seventeenth century, when it was used to throw explosives.[2] The last recorded military use of slings was during the Spanish Civil War (1936–39), when troops lobbed grenades instead of stones and lead-shot. Even in modern-day conflicts, the sling acts as a tool of violence. The Israeli-Palestine border conflicts see the regular use of slings as makeshift weapons, to such an extent that studies have been conducted on the traumatic potential of them.[3]

Throughout the Graeco-Roman period, slingers – Greek *sphendónitai* (σφενδονῆται)/Latin *Funditores* – were extensively fielded on many battlefields by a variety of empires, kingdoms and peoples. The abundant use of slingers in antiquity is made clear by the fourth-century military manual of Vegetius, *De Re Militari*: 'That slingers served in all battles of the ancients is known to everyone.'[4]

Slings were inexpensive and easy to construct, often designating them as the weapon of the poor and those of low social status. In the Servian Roman Army, for instance, slingers were deemed the lowest of the military arm and recruited from the poorest citizens who qualified to join the military as combatants.[5] By the time of the Republic, citizens were no longer armed with slings, and as with archers, the role of slingers was filled by foreign mercenaries and auxiliaries. The most famous story of the sling is undoubtedly the Biblical tale of David and Goliath.[6] Even here, the sling as the weapon of the poor, low-status herdsman is contrasted with the illustrious arms of the heavy infantry champion. As with the archer, the slinger did not meet the criteria of Graeco-Roman courage because they did not stand and fight the enemy face-to-face, instead fighting from a distance. Again, we are reminded of the Locrians in the *Iliad*, who carried slings and bows because they lacked the courage to engage the enemy in hand-to-hand fighting. Slingers thus suffered the same type of bias as archers. Perhaps one of the most telling of cases is Demosthenes' opposition to Charidemus, where he taunts the character of Charidemus by highlighting his low-class origins and that he started as a mere mercenary slinger.[7] Because of this bias against slingers, the literary and especially iconographical sources on these troops are relatively rare (**see plate 14**).

As a sidenote, despite Charidemus' humble origins and having to start his military career as a mercenary slinger, he rose far above this position by becoming a successful pirate captain, then a mercenary captain who fought both for and against Athens, attempting to carve out his own principality in Asia Minor. He also fought for and against Thrace, marrying into a Thracian noble family, being awarded Athenian citizenship, serving as *trierarch* (commander of a trireme) and eventually being elected to the position of *strategos* (general) of Athens in its war against Philip II. Charidemus became a mercenary captain once more after fleeing Athens to escape Alexander's wrath. He was executed while in Persian employ after criticizing King Darius's troop quality and preparations prior to the Battle of Issus (333 BC).

Archaeological finds of slings are also extremely rare because of the biodegradable material of the weapon. However, sling missiles in the form of moulded lead and baked clay bullets are common finds at excavated battle and siege sites.

Despite the sling's simplistic design and it being wielded by those who were considered to be second-class warriors, the effectiveness and lethality of the sling cannot be denied. Onasander states that 'the sling is the most deadly weapon used by the light troops', and of course it is David, the poor shepherd boy, who kills the heavily armoured Goliath, champion of the Philistines, with a simple sling.[8]

Slings rarely feature in Homer's *Iliad*; indeed, the Greek word for sling (*sphendónē*) in the context of a weapon appears only once in the entire epic. It is used to describe the armament of the Locrians, who 'had no brazen helms with thick plumes of horse-hair, neither round shields, nor spears of ash, but trusting in bows and well-twisted slings of sheep's wool'.[9] With their slings and bows, the Locrians took cover behind the phalanx and 'shot thick and fast, and sought to break the battalions of the Trojans'. Elsewhere in the *Iliad*, 'like snow-flakes' stone missiles 'fell ever earthward', but it is ambiguous whether they are thrown by hand or from a sling.[10] Herodotus only mentions slingers in the context of Gelon of Syracuse's offer of sending troops, among them 2,000 slingers, to aid the Greeks against the Persians.[11]

After the Persian Wars, slingers appear far more regularly in the pages of Graeco-Roman history. Slingers featured on both sides of the conflict during the Peloponnesian War. Thucydides tells us that the Acarnanians were considered the best slingers among the Greeks.[12] The Acarnanian slingers from Stratos did indeed give a good account of themselves, fighting off the Spartans and their allies who attempted to besiege their city in 429 BC. The Battle of Sphacteria (425 BC) shocked the Greek world when a Spartan hoplite army was battered into submission by the light troops of the Athenian general Demosthenes. Not only was it unthinkable that Spartiates would surrender, but the battle demonstrated the effectiveness of light troops against heavy infantry under the right circumstances. The very best soldiers Sparta could field were defeated by those of the lowliest social status and deemed inferior on the field of battle, many of whom were slingers.

Nicias, in his speech to the Athenian assembly on the preparations needed for the Sicilian expedition, highlighted the need for combined arms to take on the enemy forces. In particular, he emphasized the requirement for archers and slingers to counter the enemy's strong cavalry force.[13] For the expedition, Athens thus hired a large number of slingers, 700 in total, from the island of Rhodes who had a reputation for being excellent shots.[14] Later, Demosthenes gathered even more light troops to reinforce the Athenian forces already in Sicily, this time including the skilled Acarnanian slingers.[15] On the Peloponnesian side, we see the Boeotians having gathered slingers from the Malian Gulf in their preparations to besiege Delium (424 BC).[16]

The perilous march of the Ten Thousand under Xenophon demonstrated the effectiveness of light troops in harassment. Here again we see slingers being deployed on both sides of the conflict. In their attempt to march back to Greece, Xenophon's hoplites came under attack from 400 Persian slingers and archers and 200 cavalry:

> '[The Persians] began shooting with their bows and others with slings, and they inflicted wounds. And the Greek rear-guard, while suffering severely, could not retaliate at all; for the Cretan bowmen not only had a shorter range than the Persians, but besides, since they had no armour, they were shut in within the lines of the hoplites; and the Greek javelin-men could not throw far enough to reach the enemy's slingers.'[17]

The Greek hoplites and peltasts made several futile attempts to rush out and drive off the enemy, but the Persians were far too nimble for the Greeks to catch. The Greeks were clearly lacking in light troops to defend their vulnerable marching column. Xenophon's response was that 'we need slingers ourselves at once', and he rearmed as slingers those hoplites among them who hailed from Rhodes, along with other volunteers who had slinging experience, bringing their total number to 200. It is interesting to note that Xenophon highlights the difference between the Greek and Persian slingers. The Persians slung stones which were 'as large as the hand can hold'. The same Carduchi who used the large arrows against Xenophon's troops also flung the 'largest stones they could with slings', and with these 'missiles they inflicted wounds on the Greeks ... slaying many and seriously injuring not a few'.[18] The Rhodians, on the other hand, were 'versed also in the art of slinging leaden bullets'. This is the first mention of lead bullets or slingshot in Greek literature, but the use of lead slingshot certainly predates Xenophon's *Anabasis*. According to Xenophon, the lead bullets had twice the range of the heavy stones of the Persians and even outranged the arrows from the Persian bowmen.[19] The following day, the Persians attacked with an even larger number of light troops, but the combined efforts of the Greek slingers and archers 'sent back an answering volley, and not a man among them missed his mark (for even if he had been very eager to do so, it would not have been easy), then Tissaphernes withdrew out of range with all speed, and the other battalions followed his example'.[20] Xenophon's account of the skirmish highlights the importance of light troops to protect an army on the march. Slingers were of particular importance due to their long-range ability when using lead slingshot.

During the Corinthian War (395–387 BC), we once again see Acarnanian slingers giving a good account of themselves against the Spartan forces of Agesilaus, dislodging them 'with volleys of sling-stones and rocks' from the mountainside where the Spartans were encamped.[21] In 369 BC, at the time of the Theban hegemony, a Theban attack on Corinth resulted in a group of slingers and archers attacking the vanguard of the Theban picked hoplites, and 'with volleys of sling stones and arrows they laid low a pretty large number in the van of the attack, routing them'.[22] The skirmish was deemed significant enough for the Corinthians to set up a trophy to commemorate the victory over the Thebans. The following year, the Thessalian allies of Thebes used their javelinists and slingers to great effect in driving off the Athenian cavalry.[23] We are told by Diodorus that the Thessalians were expert slingers and javelineers because they 'practised from boyhood assiduously this type of fighting'.[24] In all three cases, we see elite hoplites and cavalry beaten back by slingers, despite them being deemed by the Greeks as inferior troops.

Both Philip and Alexander made use of slingers in their Macedonian armies. Philip used slingers in his siege of Olynthus (348 BC,) and Diodorus and Arrian mention slingers in the army of Alexander on several occasions, in his campaigns against the Persians, Scythians and Indians.[25] During the Hellenistic Age, Alexander's Successors made extensive use of slingers. Antigonus gained a brilliant victory over Cleitus through a daring night attack with his 'bowmen, slingers, and a sufficient number of other light-armed troops'.[26] When Eumenes prepared for his war with Antigonus, he had a staggering amount of archers and slingers, 10,000 in number, who joined his ranks under leadership of Peucestes.[27] Diodorus tells us that this Peucestes was the satrap of Persis and that the natives of this land were 'the most warlike of the Persians, every man being a bowman and a slinger'.[28] Curtius Rufus (5.6.18) also mentions that the Mardi, a Persian people, 'bind their brows with a sling; this is both an ornament of their heads and a weapon'. At the Battle of Ipsus (301 BC), 500 of these 'warlike' Persian slingers were dispersed, along with other light troops, between Antigonus' elephants.[29] Demetrius, the son of Antigonus, defended Syria against a potential invasion from Ptolemaic Egypt with an army which contained 400 archers and slingers.[30]

When Molon, the satrap of Media, rebelled against the Seleucid king Antiochus the Great in 221 BC, Polybius tells us that Molon planned to fight Antiochus' army in the hill country of Apollonia because 'he relied on the numbers of his force of slingers known as Cyrtii'.[31] The Cyrtii were an Iranian people, possibly identifiable with later Kurds, hailing from the Zargos mountains.[32] Molon was an experienced commander and had already defeated two Seleucid armies sent against him. For Molon to rely so much on his slingers to defeat Antiochus speaks highly of these men's skill with the sling. The battle between Antiochus and Molon was, however, not fought in the hill country but on an open plain which favoured cavalry and the heavy infantry of the phalanx, and less so the slingers of Molon. We do see Antiochus the Great fielding his own Cyrtii slingers in vast numbers against Rome at the Battle of Magnesia (190 BC). Livy groups them with Elymaean archers and numbers the Cyrtii and Elymaeans at 8,000, with 4,000 on each flank of the Seleucid army.[33] Along with Cyrtii slingers, Antiochus had Agrianian and Persian slingers whom he fielded against Ptolemaic Egypt at the Battle of Raphia (217 BC).[34] We are also told by Polybius that Antiochus' hazardous march through the pass of Mount Labus (210 BC) was only made possible by the efforts of his light troops, and that of all these light troops the 'most deadly weapons however proved to be the slings, which could carry a great distance'.[35] In 192 BC, Philopoemen, the famous Achaean general, executed a devastating amphibious night raid on the Spartan camp with his *caetrati*, who were predominantly armed 'with slings and darts'.[36] After his successful night attack, Philopoemen invaded Lacedaemon (Sparta's home territory) itself and caught the Spartan forces of Nabis in an ambush using these same *caetrati* troops. According to Livy, the Achaeans, particularly those from Aegium, Patrae and Dymae, were excellent slingers, surpassing in skill even the famed Balearic slingers:

> 'These peoples were trained from boyhood, in accordance with a tradition of the race, in hurling with a sling at the open sea the round stones which, mingled with the sand, generally strew the coasts. In consequence they

> use this weapon at longer range, with greater accuracy and with more powerful effect than the Balearic slinger. Moreover, the sling is not composed of a single strap, like those of the Baleares and other peoples, but the bullet-carrier is triple, strengthened with numerous seams, that the missile may not fly out at random, from the pliancy of the strap at the moment of discharge, but, seated firmly while being whirled, may be shot out as if from a bow-string. Having been trained to shoot through rings of moderate circumference from long distances, they would wound not merely the heads of their enemies but any part of the face at which they might have aimed.'[37]

Whether they were truly superior to the Balearics is debatable, as most sources exclaim the superiority of Balearic slingers, but there is little doubt that the Achaeans were skilled with the sling. During Rome's siege of Same (189 BC) during the Aetolian War (191–189 BC), it was Rome's Achaean slingers who were credited with breaking every sally attempt made by the Sameans with their deadly aim. They were so accurate and deadly 'to such an extent that from the walls they (the Sameans) begged the Achaeans to withdraw for a while and in quiet to watch them fighting with the Roman outguards'.[38] The Achaeans were so skilled with their slings that the phrase 'Achaean missile' was used to denote an excellent hit, a phrase akin to the modern 'bull's eye'.[39] This is perhaps why Philip V of Macedon had 300 of these Achaeans slingers to aid him in his siege of Ambracus (219 BC).[40]

Further west in the Mediterranean, slingers served extensively in the armies of Magna Graecia, Rome, Spanish Iberia, Carthage and other smaller African kingdoms. Dionysius, the tyrant of Syracuse, used slingers to great effect in a naval engagement against the Carthaginians at Motya in 398 BC. Due to the 'great number of archers and slingers', the Carthaginian ships were unable to come close to the Greek vessels because of the 'multitude of missiles' from the slingers and archers.[41] Agathocles, another tyrant of Syracuse, also had a contingent of slingers in his army during his invasion of Africa. He used his slingers to defend his army on the march from the harassing 'nomads' (possibly Numidians) in a fashion not all that different from how Xenophon used his Rhodian slingers against the Persians, and in his successful siege against Utica (307 BC).[42]

I have already mentioned that slingers formed the lowest-ranked part of the Roman Servian army. While Rome no longer fielded slingers from its citizens by the time of the mid-Republic (290–88 BC), it nevertheless recognized the value of these range troops and hired them as mercenaries or fielded them from allied or subject peoples. During the Second Punic War, Rome's ally,

Hiero II of Syracuse, provided the Romans with 1,000 archers and slingers 'capable of acting against the Balearics and Moors and other tribes who fight with missile weapons'.[43] Two years later (214 BC), Rome besieged Syracuse after Hiero broke his alliance with Rome and sided with Carthage. During the siege, Rome used slingers and archers in amphibious actions to assault the city.[44] In 207 BC, Hasdrubal arrived in Italy to reinforce his brother, Hannibal. Rome immediately enlisted as many men as possible to combat the new threat, going as far as enlisting slaves. According to Livy, 3,000 slingers and archers were sent as reinforcements from Sicily. During Rome's war with the Galatians (189 BC), the campaign took place in mountainous, uneven terrain unsuitable for heavy infantry. Preparations were thus made to sufficiently equip the light infantry, including numerous slingers with an abundance of missiles:

> 'The consul (Cn. Manlius) had anticipated that the fighting would not be at close quarters but would involve an attack upon positions from a distance, [so] he accumulated a large quantity of javelins, light infantry spears, arrows and leaden balls and small stones suitable for hurling from slings.'[45]

The Galatian tribes suffered a devastating defeat at the hands of the light infantry. However, when Manlius requested a triumphal procession for his victory over the Galatians, he was apparently mocked because his victory was not won with the sword, but with 'slings and arrows'.[46] The account goes to great lengths to emphasize with what little regard light troops were viewed.

I have already mentioned the expert Cyrtii slingers in the army of the rebel Molon and in that of Antiochus the Great. At the Battle of Callinicus (171 BC), Rome fielded 300 of these Cyrtii, who were supplied by their faithful ally Eumenes of Pergamum.[47]

During the late Republic, Sallust mentions the Romans using slingers against the Numidians in the Jugurthine War (112–106 BC).[48] When Pompey assembled his forces in Greece in preparation for his war with Caesar, he gathered six cohorts of slingers.[49] Both Caesar and Pompey fielded slingers at the Battle of Pharsalus (48 BC), and Caesar successfully used slingers and archers against Scipio's elephants at the Battle of Thapsus (46 BC).[50] In 38 BC, the Roman general Ventidius did much to redeem the humiliating defeat the Romans suffered at Carrhae by defeating a Parthian army at Cyrrhestica. During the battle, slingers played a crucial role in beating back the Parthian heavily armoured *cataphracts*:

> '[T]he majority of them [the Parthian cavalry] being in armour, yet they were confused by the unexpectedness of the onslaught and by stumbling over one another and were defeated by the heavy-armed men and especially

> by the slingers; for these struck them from a distance with their powerful missiles and so were exceedingly difficult for them to withstand.'[51]

Ventidius' successes paved the role for Antony's campaign against the Parthians (36 BC), where Roman slingers once again gave an excellent account of themselves by defeating both the Parthian horse archers and heavily armoured *cataphracts*:

> 'He [Antony] routed them [Parthians], for as his slingers were numerous and could shoot farther than the archers, they inflicted severe injury upon all, even upon the men in armour; yet he did not kill any considerable number of the enemy, because the barbarians could ride fast.'[52]

It is, however, among the armies of Carthage, Rome's great nemesis, that we meet antiquity's most famed and skilled slingers, those who inhabited the Balearic Islands. As with Cretan archers, a special section (see Chapter 6: Balearic Slingers) is dedicated to these islanders, who with their slings 'were the most skilled of all men'.[53]

3.1 Arms and Armour

3.1.1 The Sling

The sling was a very cost-effective weapon and relatively easy to construct. Slings were constructed from different materials and came in different lengths, and while variable in detail, the weapon shared the same basic principles across cultures and time. Its design was simplistic, being a pouch or pocket between two cords. At the end of one cord was a loop which fitted around the middle finger (or at times around the wrist); at the end of the other cord was a knot which was held between the thumb and index finger.

According to Strabo (3.5.1), the slings used by the Balearic islanders were constructed from 'black tufted rush ... hair, or sinews'. According to Vegetius, slings were constructed from linen or 'thick bristly hair' (*saetae*).[54] On the opposite spectrums of time and space, the sling found in the tomb of King Tutankhamun (1352 BC) was plaited from linen, while a sling discovered in a ninth-century AD Tibetan fortress was woven from goat's hair, corresponding well with Vegetius' description of them.[55] Leather, hemp, flax or the 'well-twisted wool' mentioned in the *Iliad* were also used as materials for sling construction.[56] Xenophon writes that the slings used by the Rhodians were plaited, while Livy describes the slings of the Achaeans as consisting of 'three thongs, stiffened by being sewn together', as opposed to the Balearic sling, which was 'made of a single strap'.[57] While Livy's description is somewhat unclear, it seems to suggest that the cords were constructed separately and then attached to the pocket by a method of stitching. The pocket of the sling was at times constructed from a different material than the cords. The pocket could, for instance, be woven from wool or made of leather, whereas the cords were plaited from flax **(see plate 15)**. Livy's description of the Balearic slings being 'made of a single strap' is also interesting to note, as the sling, the cords and pocket could be plaited from a single piece of material. In these cases, the pocket is often split, creating a gap or hole within the pocket in which the missile would be placed **(see plate 16)**. Within modern slinging communities, this type of sling is still referred to as a 'Balearic sling' or a 'split pouch sling'. Slings could also be of different lengths and sizes, allowing them to be used for different distances, and for smaller or larger slingshot missiles.

3.1.2 Slingshot

Stones

A variety of types of slingshot could be used as missiles for the sling, the most basic being a simple stone. We are reminded once again of the story of David, the shepherd boy who collected 'five smooth stones from the brook', one of which he shot from his sling to slay the giant Goliath.[58] As mentioned above, Livy attributed the Achaeans' skill with the sling to the fact that they 'were trained from boyhood, in accordance with a tradition of the race, in hurling with a sling at the open sea the round stones which, mingled with the sand, generally strew the coasts'. This is perhaps the very same reason why the Rhodians were so skilled at slinging, as the shores of the island are covered with smooth stones. These smooth river and beach stones had an obvious aerodynamic advantage over rough-edged stones. Shaped stones as sling missiles start to appear as early as the sixth millennia BC.[59] Limestone was naturally a popular material for manufactured stone slingshot, allowing for the stones to be shaped to be not only as aerodynamic as possible, but also to be standardized regarding weight, shape and size to maximize accuracy and effectiveness.

As we have seen, Xenophon wrote that the Persians slung stones 'as large as the hand can hold'.[60] Similarly, we are told that the Balearics slung large stones weighing a *minia* (approximately 350–450 grams).[61] Vegetius, in a much later period (fifth century AD), wrote extensively on the sling but only mentioned the use of stones as opposed to other types of sling missiles. Vegetius also highlighted that battlefields would often naturally be dotted with stones, allowing slingers to replenish their ammunition.[62]

Clay slingshot

The advantages which clay held in shaping slingshot missiles were recognized at a very early stage, and clay slingshot can be found dating as far back as 7000 BC.[63] It is interesting to note that in most cases clay slingshot was sundried as opposed to oven-baked. In order for the clay to retain its density and the maximum weight for the limited dimensions of the bullet, the clay was not tempered with chaff, as was done with conventional brick or pot making. Without the chaff, the clay slingshot would have cracked in the oven fire. On very few occasions, stone pebbles were packed inside the clay slingshot.

During Scipio's lengthy siege of Carthage (149–146 BC), slingers fought on both sides of the conflict. A copious amount of slingshot was discovered during excavations of Byrsa Hill, the citadel of the capital. A large amount of the slingshot discovered at the site was clay slingshot, which is currently on display in the Byrsa Hill Museum (**see plate 17**).

Interestingly, Caesar mentions in his *Gallic Wars* (5.43.1) that slingers made use of heated clay slingshot to set thatched huts alight:

> 'On the seventh day of the attack, a very high wind having sprung up, they began to discharge by their slings hot balls made of burned or hardened clay, and heated javelins, upon the huts, which, after the Gallic custom, were thatched with straw. These quickly took fire, and by the violence of the wind, scattered their flames in every part of the camp.'

Lead slingshot

During the Classical Greek period, lead slingshot became commonly used for sling missiles. As we have seen, the first literary account of lead slingshot appears in Xenophon's *Anabasis*, where the Rhodians were 'versed also in the art of slinging leaden bullets'. The use of lead slingshot, however, certainly predates Xenophon's account.

Because of the density of lead, the slingshot projectile retained a maximum weight in a small volume, and due to its small size had minimum air resistance. Lead slingshot was generally made by pouring molten lead into two-part moulds and being left to cool. This helped tremendously with the standardization of slingshot, maximizing its efficiency. Lead slingshot came in a variety of shapes and sizes, an ellipsoidal or 'acorn' shape being particularly popular. The Latin name for lead slingshot, *glandes plumbeae*, literally meaning 'leaden acorns', reflecting the 'acorn' shape. Another popular design was a spherical or biconical almond shape. Impromptu lead slingshot could even be prepared on the battlefield by pouring molten lead into simple holes in the sand, as attested by Tacitus.[64] Archaeological evidence of such improvised slingshot casting was also discovered at Velsen Fort in the Netherlands.[65]

Because of the heavy weight and relatively small size of the missile, as well as its aerodynamic features, slingers using lead slingshot could easily outrange those using stones or even archers. Xenophon states that the Rhodian slingers' 'missile carries no less than twice as far as those from the Persian slings', and 'farther even than the Persian bowmen'.[66] Cassius Dio also confirms that slingers outranged archers.[67] Onasander highlights that 'the lead slug is the same colour as the air and invisible in its course, so that it falls unexpectedly on the unprotected bodies of the enemy'.[68]

One of the most fascinating things about lead slingshot missiles was that they were often cast with inscriptions upon them. Of the nearly 500 lead slingshot discovered at Olynthus, 112 were inscribed. These inscriptions are of the utmost interest as they give a glimpse into the past. They are often inscribed with personal names, usually those of squadron commanders, or identification of a specific unit, ethnicity or civic allegiance.[69] They are also a source of information on military humour, wit and a sense of the martial psyche. Numerous slingshot have inscriptions which denote messages to their enemies, such as 'fire', 'it rains', 'ouch', 'take that', 'crack your teeth' and even 'attack Pompey's arsehole' (**see plate 18**).[70]

3.1.3 Kestros/Cestrus

During the Third Macedonian War (171–168 BC), a new type of weapon appeared on the Roman battlefield, the *kestros* (*κέστρος*) or *kestrophendone*, Latinized as *cestrus* or *cestrosphendone*. As the name indicates, it was a type of sling. Both Polybius and Livy describe the *kestros* as a novel weapon, being a sling which flung large darts, which was used by the Macedonians during the war. According to Polybius, these darts consisted of a metal head and tube two palms (152mm) long and a wooden shaft a span (228mm) in length and a digit (19mm) in diameter:

> 'The *kestros* was a novel invention, made during the war with Perseus. This weapon consisted of an iron bolt two palms long, half of which was spike, and half a tube for the reception of the wooden shaft which was fixed into the tube, and measured a span in length and a finger-breadth in diameter. At the middle point of the shaft three wooden "plumes" were morticed in. The sling had thongs of unequal length, and on the leather between them the missile was loosely set. When the sling was being swung round, with the two thongs taut, the missile kept its place; but when the slinger let go one of the thongs, it flew from the leather like a leaden bullet, and

> was projected from the sling with such force as to inflict a very grievous wound upon any one whom it hit.'[71]

Livy's description of the *kestros* mimics that of Polybius, but he does provide a more detailed account of the effectiveness of the weapon:

> 'The Romans were in very great danger, for they could not fight in close order against those who were struggling up the hill, and if they left their ranks and ran forward they were exposed to the javelins and arrows. They suffered mainly from the *cestrosphendons*, a novel kind of weapon invented during the war. It consisted of a pointed iron head two palms long, fastened to a shaft made of pinewood, nine inches long and as thick as a man's finger. Round the shaft three feathers were fastened as in the case of arrows, and the sling was held by two thongs, one shorter than the other. When the missile was poised in the centre of the sling, the slinger whirled it round with great force and it flew out like a leaden bullet. Many of the soldiers were wounded by these and by missiles of all kinds, and they were becoming so exhausted that they were hardly capable of holding their weapons.'[72]

While this is the only mention of the weapon in antiquity, and the *kestros* never became a popular battlefield weapon, from the above account it does appear that, at least in this one instance, it gave a good account of itself.

Experiments with working reconstructions of the *kestros* sling date back to 1874. These early efforts were able to sling the dart an impressive distance of 70 metres.[73] A more recent reproduction which corresponded more closely to Polybius' description was able to sling the dart a staggering 97 metres.[74]

3.1.4 Lithoboloi

The simple action of throwing a stone by hand could be deadly, and literary sources attest to stones being thrown by hand on ancient battlefields. In the *Iliad*, numerous characters throw stones by hand to smite their foes. Agamemnon rages among the ranks of the Trojans with spear and shield, but also with rocks that he threw at the enemy.[75] Aeneas attempted to throw a huge stone at Achilles, but Poseidon whisked Aeneas from the field of battle before he could do so.[76] Hector killed Epeigeus with a stone and 'his head was wholly cloven asunder within the heavy helmet'.[77] Hector's charioteer is also killed by Patroclus with a well-placed stone:

'[Patroclus] grasped a stone, shining and jagged, that his hand compassed about. Firmly he planted himself, and hurled it, neither had he long awe of his foe, nor sped he his missile in vain, but smote the charioteer of Hector, even Cebriones, a bastard son of glorious Priam, upon the forehead with the sharp stone, as he was holding the reins of the horses.'

Even the immortal gods did not refrain from throwing stones by hand, Athena striking down blood-stained Ares with 'a stone that lay upon the plain, black and jagged and great'.[78]

While none of the above-mentioned characters would be considered light troops, it does demonstrate the effectiveness of a simple stone or rock thrown by hand. Actual stone-throwers, or *lithoboloi*, as a type of light troop unit whose primary function on the battlefield was to throw rocks at the enemy, are well documented in ancient literature. This role was often filled by camp followers and attendants, porters, civilians caught in a siege or simple oarsmen who took part in amphibious assaults. These impromptu 'soldiers' could be used to swell the ranks, the simple action of throwing rocks needing little to no training or equipment other than the collecting of stones.

Plato states in his *Laws* that tournaments should be held where light troops competed in all types of range weapons, including 'stones flung either by hand or by sling', to help them train for war.[79] Eight centuries later, Vegetius (1.16) still advocates that troops should 'be thoroughly trained at casting stones by hand or with slings'.

Thyere are numerous recorded instances of Greek forces throwing stones in combat, or coming under fire from such weapons. During the Peloponnesian War, a contingent of Corinthian hoplites was trapped in a defile and stoned to death by Athenian light troops (*psiloi*).[80] At the Battle of Sphacteria (425 BC), stone throwers are mentioned separately from the slingers.[81] It is very likely that these stone throwers came from the ranks of the 7,000 rowers who accompanied Demosthenes on his campaign. In 415 BC, during the disastrous Athenian campaign in Sicily, stone throwers are again mentioned separately from the slingers, fighting on both sides of the conflict.[82] The Athenian general Thrasybulus, who led the democratic resistance against the Thirty Tyrants, successfully drove off a Spartan hoplite army with his light troops, which included men who threw stones by hand, in 403 BC.[83] Finally, Alexander the Great and his soldiers had to storm the Susian Gates (330 BC) under a hail of stones thrown by hand from the enemy.[84]

At the Battle of Baecula (209 BC) during the Second Punic War, both Roman infantry and camp attendants resorted to throwing 'stones which the place affords, strewn broadcast and almost all of them of a size to be thrown' in response to the Carthaginian light troops. Stone throwers are also mentioned by Josephus during the Roman Imperial period fighting alongside Syrian slingers and Arabian archers at the siege of Jotapata (AD 67).[85]

Needless to say, such lowly irregular combatants were not a heroic topic for artists and so rarely feature in contemporary art. There are, however, a few representations of stone-throwers in art. The Column of Trajan serves as such an example, with depictions of irregulars who throw rocks by hand, fighting alongside slingers and legionaries.

Perhaps the most significant 'stone throw' in history came from the most unlikely of people. The great Pyrrhus of Epirus died an inglorious death at the Battle of Argos (272 BC) when he was struck by a roof tile thrown by an old woman, the mother of the man whom Pyrrhus was about to slay:

> '[The old woman] lifting up a tile with both her hands threw it at Pyrrhus. It fell upon his head below his helmet and crushed the vertebrae at the base of his neck, so that his sight was blurred and his hands dropped the reins. Then he sank down from his horse and fell near the tomb of Licymnius.'[86]

Another prominent historical figure who lost his life due to a stone thrown by hand was the Persian military leader Mardonius at the Battle of Platea (479 BC), when he 'was slain by a man of Sparta named Arimnestus, who crushed his head with a stone'.[87] Even the lowliest and most rudimentary of weapons could change history!

3.2 Slingers and Stone-throwers in Combat

The inexpensive and easy construction of a sling, as well as its size, made it possible for almost any man to arm himself with such a weapon. The *De Re Militari* of Vegetius recommended that every Roman soldier 'without exception' should be instructed in the use of the sling, for it 'cannot be reckoned any incumbrance, and often is of the greatest service'.[88]

Range Fighting

The primary function of slingers was to attack the enemy from afar. Indeed, one of the greatest advantages of slingers was their exceptional range ability. Those who used lead slingshot and moderate clay slingshot or stones enjoyed a huge range advantage over other light troops such as javelinists and even archers. I have already mentioned Xenophon's Rhodians, who used lead slingshot which carried 'no less than twice as far as those from the Persian slings' and 'further even than the Persian bowmen'. The perilous journey back home of the Ten Thousand was only possible due to the effectiveness of the Rhodian slingers. Prior to the deployment of the Rhodian slingers, the Ten Thousand suffered severely under the missile harassment of the Persians. The Rhodians could keep the Persian missile units back due to their superior range, ensuring the safety of the marching column. At the Battle of Mount Labus (210 BC), the Seleucids gained a decisive victory over the 'barbarians' through effective use of their light troops. Polybius (10.30.7–8) states that out of all the light troops, the 'most deadly weapons however proved to be the slings, which could carry a great distance'. It was this range advantage of slingers which allowed the Romans in the armies of Ventidius and Mark Antony to counter the horse archers of the Parthians, because 'slingers were numerous and could shoot further than the archers'.[89]

The exact range of ancient slingers is unfortunately not recorded. Several modern studies have been conducted on how far a slingshot could be hurled, with varied results.[90] Obviously, these experiments depended on numerous elements, such as what type of material the slingshot was made of, the weight of the missile, the size of the sling, the experience of the slinger and the style of slinging method, as well as the impact of natural elements such as wind upon the missile. A vibrant international slinging community still exists, with international competitions and a very active web forum dealing with almost every element of slinging, from history and archaeology to sling construction and different slinging styles.[91] Extraordinary distances of well over 400 metres have been reported by community members.[92] While it might be impossible to determine the effective range of ancient slings, there is no doubt that slingers could cast their missiles over extraordinary distances.

Blunt Force Trauma

The sling and stones thrown by hand also had an advantage over other missile weapons when it came to armour. Slingshot and large stones did not need to penetrate skin or tissue to inflict wounds, as arrows or javelins did. The forceful impact of these missiles could transfer through the armour into the body and had the ability to cause blunt force trauma under the armour, with severe contusions, abrasions, lacerations, internal haemorrhages and bone fractures. This was particularly true for the 'fist-size' stones mentioned by Xenophon or the *mina*-size stones used by the Balearics. The force produced upon impact was often enough to crush or break defensive armour. I have already mentioned Cassius Dio's accounts of the effects of Roman slingers upon the armour of Parthian cataphracts, but they are worthy of repetition here:

> '[The armoured Parthian cataphracts were defeated] especially by the slingers; for these struck them from a distance with their powerful missiles and so were exceedingly difficult for them to withstand.'[93]
>
> '[Mark Antony's] slingers were numerous and could shoot farther than the archers, [and] they inflicted severe injury upon all, even upon the men in armour.'

Vegetius confirms the advantage slingshot held over arrows when it came to armour:

> 'Often, against soldiers armed with helmets, cataphracts and cuirasses, smooth stones shot with a sling or "sling-staff" are more dangerous than any arrows, since while leaving limbs intact they inflict a wound that is still lethal, and the enemy dies from the blow of the stone without loss of blood.'[94]

Experimental archaeology has also shown that a lead slingshot weighing a mere 45 grams had the potential to crush period bronze helmets, as demonstrated in Chapter 8 (**see plate 20**).[95]

The devastating effects of blunt force trauma have been demonstrated through experimental archaeology. In an experimental test, a stone slingshot weighing 120 grams was cast with a traditional sling braided from hemp and wool, at a pig skull fully intact with flesh and skin. The frontal bone and a section of the zygomatic arch of the skull was completely shattered upon impact (**see plate 21**). While porcine skulls are very similar to human skulls and have been used as a substitute for the latter, they are more robust and have a slightly higher bone density. The porcine epidermis, too, while comparable to human skin, is higher in density, thickness and strength. The trauma caused by the stone projectile upon a human head would thus have been far more severe.[96]

Penetrating Trauma

There is, however, literary evidence that slingshot could penetrate the skin and lodge itself in the tissue of the body, creating an open wound. Slingshot that had this capability were obviously smaller, bi-pointed missiles of lead, clay or shaped stone. Onasander states:

> '[N]ot only is the impact itself violent, but also the missile, heated by the friction of its rush through the air, penetrates the flesh very rapidly so that it even becomes invisible and the swelling quickly closes over it.'[97]

First-century physician Celsus, in his medical treatise *De Medicina,* also mentions the ability of slingshot to embed itself into the flesh of the victim and gives a description of how it should be removed:

> '[T]here is a third type of [projectile] that sometimes needs to be removed, a leaden bullet or rock or something similar, which breaking through the skin lodges inside in one piece. In all of these cases, the wound needs to be opened a bit wider, and what is inside must be extracted with pincers along the same pathway by which it entered.'[98]

On the other side of the world, a recent archaeological find in Guam produced the skull of a man which was shattered by a bi-pointed stone slingshot which was still embedded within. While far removed from the Classical Greaco-Roman world in both space and time, it nonetheless provides evidence of the devastating penetrative capabilities of slingshot.[99]

In contrast to most sling missiles, the primary function of the *kestros*, using a dart as a missile, was to penetrate. From Livy's account, it proved to be fairly effective as the Romans 'suffered mainly from the *cestrosphendons*' rather than from other missiles such as arrows and javelins.

Quantity of Ammunition

The relatively small size of lead and clay slingshot, as well as the smaller stones, allowed for a considerable number of missiles to be carried. Javelinists could carry approximately seven javelins, while infantry archers carried quivers that seemed to hold a maximum of sixty arrows. It was also far easier to produce lead or clay slingshot than javelins or arrows. As mentioned, clay slingshot was simply sun-baked, while molten lead could be easily poured into a premade cast. Under more pressing circumstances, lead slingshot could even be made by pouring lead into simple holes in the sand. Arrows, however, had to be fletched with feathers, the reed shafts straightened by steaming and the arrowheads cast in bronze or forged if made of iron, an even more complex task. It was thus far easier and cheaper to produce ammunition for slings. Even when slingers did run out of ammunition, they would often be able to simply collect stones from the battlefield to replenish their stock as, in the words of Vegetius, 'it often happens too that warfare is carried on in stony places'.[100]

Siege Warfare

Like archers, slingers and those who threw stones by hand played an important role in siege warfare, being used by both the besieged and the besiegers. The *Iliad* has several references to the use of stones and rocks in the defence of fortified positions, as well as in attacking them. Book 12 of Homer's epic recounts the Trojan attack on the Grecian encampment wall, with stones hurled from both sides:

> 'For the men above kept hurling stones from the well-built towers, in defence of their own lives and of the huts and of the swift-faring ships. And like snow-flakes the stones fell ever earthward, like flakes that a blustering wind, as it driveth the shadowy clouds, sheddeth thick and fast upon the bounteous earth; even so flowed the missiles from the hands of these, of Achaeans alike and Trojans; and helms rang harshly and bossed shields, as they were smitten with great stones.'[101]

Stones are again mentioned in the battle for the wall; they 'flew thick, some upon the Trojans, and some from the Trojans upon the Achaeans, as they cast at one another; and over all the wall the din arose'.[102]

I have already mentioned the Acarnanians who used their slings in defence of their city Stratos against the Spartan siege in 429 BC. Thucydides writes that these Acarnanian slingers were so effective that no man dared to leave the siege camp without wearing his armour.[103] Thucydides also mentions that the Boeotians, despite having a large contingent of light troops, sent for slingers and javelineers from the Malian Gulf specifically to assist in the siege of the Athenian fortification.[104]

The Carthaginian siege towers at the investment of Selinus in 409 BC were filled with slingers, and when the walls were finally breached and the fighting spilled over into the streets, the citizens put up a fierce resistance by throwing stones at their assailants from their rooftops.[105] Alexander's siege of Tyre (332 BC) saw the use of slingers by both the defending Tyrians and the attacking Macedonians.[106] Polyperchon, the short-lived regent of Macedonia, also used slingers in his unsuccessful siege of Megalopolis in 317 BC.[107]

In 293 BC, the Roman assault on the walls of the Samnite city of Aquilonia was repulsed by a 'shower of stones' from its defenders.[108] During the siege of Syracuse (214 BC), Rome's slingers and other light troops attacked the battlements with such ferocity that it 'allowed scarce any person to remain upon the wall unwounded'.[109] The skill and effectiveness of Rome's allied Achaean slingers during the siege of Same (189 BC) have already been discussed. At the siege of Piraeus (87–86 BC) during the First Mithridatic War (89–85 BC), slingers and other light troops were used in the defence of the walls as well as sallying out from the defences to attack the besieging Romans.[110] It is also interesting to note that during the siege, turncoat slaves sent secret messages to the Romans enclosed 'in leaden balls, and threw them over to the Romans with slings'.[111]

Naval Warfare

While slingers did not feature as prominently in naval warfare as archers, there are a few cases of slingers being used on decks of military ships. Slingers are mentioned by Thucydides as fighting on the decks of the triremes alongside archers and javelinists in the final sea engagement during the Athenian siege of Syracuse (413 BC).[112] The Syracusans also deployed slingers and archers on the decks of their ships and successfully used them to keep a numerically superior Carthaginian fleet from engaging them in 397 BC.[113] Slingers are mentioned too among the crew of the fleet of the Syracusan tyrant Agathocles in 310 BC.[114] The Tyrians had slingers stationed on their assault ships to attack Alexander's soldiers and engineers building the mole during the siege of the city in 332 BC.[115] The Roman fleet which attacked Syracuse in 212 BC contained slingers alongside other light troops.[116] Philopoemen's amphibious assault on the Spartan camp in 192 BC consisted primarily of slingers and javelinists.[117] Florus, the first-century Roman writer, mentions the boldness of Balearic pirates sailing out to sea to attack ships with their slings; they even assaulted Roman military vessels and 'at the first onslaught covered it with a shower of stones and rocks'.[118] Finally, while it was not quite a full-on naval engagement, Balearic slingers were able to prevent a Carthaginian fleet from beaching on their island with 'a shower of stones'.[119]

3.3 Conclusion on Slingers and Stone Throwers

The stone, although the most rudimentary of weapons, was very effective. While it was considered the weapon of the lowliest of combatants, the higher classes did not hesitate to pick up a stone to smite an enemy; as we saw in the *Iliad*, the most noble of heroes did not deem themselves above the action of throwing stones. Simple, yet highly effective, the weight of stone had the potential of crushing armour and breaking bones with its force of impact, with the added advantage of a range attack. This simplest of weapons thrown by hand could also prove to be fatal, as attested in the *Iliad* and proved by the stone thrown by the Spartan Arimnestus which snuffed away the life of Persian commander Mardonius at Platea. Since the stone as a missile required no construction and the act of throwing a stone needed minimal training, the ranks of armies could be swelled with camp attendants, slaves and even women who could become deadly warriors, as the death of Pyrrhus proved.

Slingers proved to be highly effective light troops, contributing tremendously to armed forces of the ancient world. This was in many ways because of the superior range of the sling, which outranged all other weapons used by light troops. The sling also served as an anti-armour range weapon. When flung from a sling, the missile travelled at such a high velocity that even a light lead slingshot of 45 grams proved very effective against armour, while larger and heavier stones – those the size of a fist, for example – hit their targets with such staggering kinetic energy that they simply crushed armour under their weight. Heavier missiles obviously sacrificed range, but compensated for it with the high kinetic energy with which they struck their targets. The combination of range and anti-armour missiles made slingers tremendously effective light infantry. Slingers were also the least encumbered of all light troops, 'because it is no effort to carry a sling'.[120] Furthermore, due to the size of lead bullets and other moderate-scale slingshot, slingers could carry a greater amount of ammunition than other light troops. Even when their ammunition was spent, slingers could resort to slinging stones which were found strewn across the battlefield. Slingers were also the most cost-effective light troops, their slings being constructed from cheap, readily available materials and their ammunition being far less expensive and easier to make than arrows and javelins. From this we can better understand Onasander's statement that 'the sling is the most deadly weapon used by the light troops'.

Chapter 4

The Javelin and the Peltast

Akontistai

Akontistai is a general Greek term used to describe javelinists. It is derived from the term *akōntion*, meaning javelin. The term was used by most Greek historiographers in all the major historical sources that used Greek as a medium when describing light troops armed with a variety of different javelins. It appeared as early as the *Iliad* and was still in use in tenth and eleventh-century AD Byzantine military manuals to describe javelin-armed light troops.[1] The term is thus very long-lived, being used throughout ancient literature.

The term *akontistai* is used in both Homer's *Iliad* and *Odyssey* for javelinmen.[2] Herodotus describes in his *Persian Wars* how, at the Battle of Salamis (480 BC), an Aeginetan ship rammed a Samothracian trireme. The Samothracian marines, being javelin-throwers (*akontistai*), used a storm of javelins to clear the deck of the Aeginetan soldiers and, while their own ship sank, crossed over to the enemy vessel and commandeered it:

> '[A]n Aeginetan ship bore down and sank the Samothracian ship, but the Samothracians, being javelin-throwers (*akontistai*), by pelting them with missiles knocked the fighters off the ship that had sunk theirs and boarded and seized it.'[3]

Akontistai featured extensively in Thucydides' *The Peloponnesian War*. During the naval battle off the Bay of Cheimerion (433 or 432 BC), *akontistai* again appear on the decks of the triremes, fighting alongside heavily armoured hoplites and archers on both sides of the engagement. Thucydides (3.97.2) refers to the Locrian *akontistai* as *psiloi*, laying emphasis on the fact that they are light troops. The Aetolian *psiloi* javelin-throwers who defeated Demosthenes's heavy-armed hoplites in 426 BC, while not directly referred to as *akontistai*, would undoubtedly fall under such a banner since they are described as *psiloi* and throwing javelins.[4]

At the Battle of Olpae (426 BC), the Athenian general Demosthenes faced a numerically superior force from the Peloponnese and was forced to extend his line by deploying his Aphilochian *akontistai* on his extreme left flank.[5] Before the battle, Demosthenes placed a contingent of '*psiloi*' and 400 hoplites in a wooded area in order to spring an ambush and attack the numerically superior Peloponnesian forces in the rear. There is little doubt that a number of these '*psiloi*' would be *akontistai*. The Amphilochian *akontistai* in the army of Demosthenes that were present at Olpae later staged an ambush against a force of Ambraciot hoplites, slaughtering a great number of them. The disadvantage of the hoplites' 'heavy armour as against light-armed' *akontistai* in the uneven terrain is heavily emphasized by Thucydides.[6]

The Boeotians, in their preparation for their attack on the Athenian fortified position at the temple of Apollo at Delium in 424 BC, specifically sent for additional reinforcements of *akontistai* and *sphendónitai* (slingers), despite having a large number of light troops of their own.[7]

It is during Athens' disastrous Sicilian expedition (415–413 BC) that we see an increased focus upon *akontistai* in Thucydides' account, so much so that it is worth dedicating some time to some of the events of the campaign. When the Athenian general and statesman Nicias describes the Syracusan forces to the Athenian assembly, he mentions that besides having large numbers of hoplites, Syracuse also has numerous *toxotai* and *akontistai*.[8] In the winter of 415 BC, the Syracusans and Athenians met in pitched battle and we are told by Thucydides that the entire right wing of the Syracusan army consisted of cavalry and *akontistai*.[9] This pairing of horsemen and *akontistai* was a recurring theme of the Syracusan military. In the summer of the following year, the Syracusan army, now led by the Spartan general Gylippus, once again met the Athenians in the field. The Syracusan army came off the worst, with Gylippus placing the blame upon himself, saying that he 'deprived them of the benefit of their cavalry and *akontistai*' by not deploying the troops in a more open field.[10] When Gylippus led the Syracusans out a second time, he deployed his cavalry and *akontistai* on the Athenian flank in the open space, where they could take full advantage of their fighting style. On this occasion, the *akontistai* and cavalry were indeed able to turn the tide and the Syracusans gained a victory over the Athenians.[11]

The increasing importance of *akontistai* during the Peloponnesian War is reflected in Thucydides' account of Demosthenes' recruitment of numerous *akontistai* to reinforce the Athenian troops in Sicily. As reinforcements, Demosthenes had gathered an unknown number of Acarnanians (7.31.5), 150 Iapygians (7.33.4), 300 from Metapontum (7.33.5) and 300 from Thurii (7.35.1), all armed as *akontistai*. Likewise, Syracuse received 300 *akontistai* from the Camarinaeans and another 400 from the Geloans (7.33.1). In 413 BC, during the sea battle in the Great Harbour, the Syracusan *akontistai* played a prominent role in the naval engagement and contributed immensely to the Syracusan victory:

> 'The Syracusans received them, and charging prow to prow as they had intended, stove in a great part of the Athenian foreships by the strength of their beaks; the *akontistai* on the decks also did great damage to the Athenians, but still greater damage was done by the Syracusans who went about in small boats, ran in upon the oars of the Athenian galleys, and sailed against their sides, and discharged from thence their javelins upon the sailors.'[12]

In the final days of the expedition, during the Athenians' desperate attempt to escape, the retreating forces were constantly attacked by Syracusan *akontistai* and cavalry, which severely impeded them and eventually ground their withdrawal to a halt:

> '[H]arassed by the charges of the Syracusan horse and by the missiles of their light troops … the Athenians advancing found themselves impeded by the missiles and charges of the horse and darters [*akontistai*], both very numerous, of the Syracusans and allies; and after fighting for a long while, at length retired to the same camp, where they had no longer provisions as before, it being impossible to leave their position.'[13]

The following day, the Athenians attempted to force a passage through a fortified pass in a bid to escape, 'but were greeted by a storm of missiles from the hill which was occupied by *akontistai*, which told with the greater effect through its being a steep one, and unable to force the passage, retreated' to their previous camp.

The next day, the army continued their retreat at a slow pace, but a large Athenian contingent under Demosthenes was eventually surrounded by *akontistai* and other light troops and 'missiles were showered in upon them from every quarter'.[14] Thucydides states that 'this mode of attack the Syracusans had with good reason adopted in preference to fighting at close quarters', emphasizing the effectiveness of light troops and skirmishing tactics against heavy-armoured troops.[15] The Athenians simply had no answer for these hit-and-run tactics and 'were worn out with their wounds and other sufferings', to such an extent that they were forced to surrender.[16]

The next day, Nicias put what remained of the Athenian 'army in motion, pressed, as before, by the Syracusans and their allies, pelted from every side by their missiles, and struck down by their javelins'.[17] They were eventually surrounded and were pelted with javelins and other missiles, which continued until the evening. At daybreak, the Athenians desperately pushed on and attempted to cross the Assinarus River (the modern Tellaro) in the hope of escaping their pursuers. The crossing, however, turned into a disaster as many fell victim to the javelins of the *akontistai*:

> '[T]he attacks of the enemy [made] it difficult to cross at all; forced to huddle together, they fell against and trod down one another, some dying immediately upon the javelins, others getting entangled together and stumbling over the articles of baggage, without being able to rise again. Meanwhile the opposite bank, which was steep, was lined by the Syracusans, who showered missiles down upon the Athenians.'[18]

The defeat at the Assinarus River (413 BC) brought an end to the Sicilian expedition and signified the beginning of the end of the Athenian Empire. The *akontistai* contributed tremendously to the victory of Syracuse and defeat of Athens. The Sicilian disaster highlights the importance the *akontistai* played in the Syracusan army and the effectiveness of their tactics, especially in difficult terrain and against heavy infantry who had no light-troop support of their own.

While Xenophon generally uses collective nouns such as *psiloi* and *gymnētēs* to indicate light troops, the term *akontistai* does appear in the *Anabasis* on several occasions to refer to javelin-armed light troops. Xenophon, for instance, highlights the Greek *akontistai* being outranged by the Persian slingers.[19] On other occasions, *akontistai* are used in conjunction with archers to attack the Carduchians.[20] The term appears more in Xenophon's *Cyropaedia*, where it is used to describe the Persian javelinmen, often fighting alongside other light troops such as slingers and archers.[21]

In Book 2 of the *Hellenica*, Xenophon differentiates between the *peltastai* (see Chapter 4: Classical Peltast) and the *akontistai*, whom he describes as light-armed *psiloi*.[22] Xenophon, however, is rather liberal with his use of terminology, and at time uses *akontistai* and *peltastai* interchangeably. For instance, in Book 6 of the *Hellenica* he states that the region around Thessaly was known for its '*akontistai*, so that it is likely that our force would be superior in *peltasts*'.[23] Similarly, in Book 3 the *akontistai* referred to are obviously the same men he calls *peltastai* a little earlier in the passage:

> '[H]e also offered prizes to *peltasts* and bowmen, for all who should prove themselves best in their respective duties. Thereupon one might have seen all the gymnasia full of men exercising, the hippodrome full of riders, and the *akontistai* and bowmen practising.'[24]

Again, the *akontistai* in the *Cyropaedia* are regarded by Xenophon as *peltastai*.[25]

Akontistai appear several times in Polybius' *Histories* to refer to javelinmen from a variety of different cultures. The term first appears in his account of the Battle of Telamon (225 BC), where the Roman javelin-armed light troops are described as *akontistai*:

> 'But when the javelineers [*akontistai*] advanced, as is their usage, from the ranks of the Roman legions and began to hurl their javelins in well-aimed volleys, the Celts in the rear ranks indeed were well protected by their trousers and cloaks, but it fell out far otherwise than they had expected with the naked men in front, and they found themselves in a very difficult and helpless predicament. For the Gaulish shield does not cover the whole body; so that their nakedness was a disadvantage, and the bigger they were the better chance had the missiles of going home. At length, unable to drive off the javelineers owing to the distance and the hail of javelins, and reduced to the utmost distress and perplexity, some of them, in their impotent rage, rushed wildly on the enemy and sacrificed their lives, while others, retreating step by step on the ranks of their comrades, threw them into disorder by their display of faint-heartedness. Thus was the spirit of the Gaesatae broken down by the javelineers; but the main body of the Insubres, Boii, and Taurisci, once the javelineers had withdrawn into the ranks and the Roman maniples attacked them, met the enemy and kept up a stubborn hand-to-hand combat.'[26]

From the above account, we can see that the deployment, tactics and armament used by these *akontistai* are the same as the *velites*, the citizen light infantry unit of the middle Republican period. There is very little doubt among scholars that these *akontistai* referred to by Polybius at Telamon are indeed Roman *velites*. Elsewhere, Roman *akontistai* were used to reconnoitre enemy movements and as support troops for a cavalry skirmish.[27] Both these roles are typically associated with the Republican *velites* and again, in all likelihood, refer to such troops. At the Battle of the River Trebia (218 BC), the *velites* appear once more, but this time as *pezakontistai* (πεζακοντιστής), which roughly translates as 'javelinmen on foot'.[28] They again fill the role of pre-battle skirmishing and supporting Roman cavalry. At the Battle of Cannae (216 BC), the *velites* are again referred to as *akontistai*, deployed as a skirmishing screening force in front of the legion and acting in support of the heavy infantry. As mentioned above, Polybius used several words to refer to Roman *velites*, including *euzōnoi*, and *akontistai* serves as yet another term to denote them.

The term in the writing of Polybius is, however, not restricted to Roman *velites*, and we see Cardacian and Lydian *akontistai* in the armies of the Seleucids at the

Battle of Raphia (217 BC).[29] It is interesting to note that in Aeschylus's extant play *Persians* (line 52), the Lydians who fought alongside the Achaemenids also acted as *akontistai*. The term again appears in a description of the forces of Antiochus at the Battle of Mount Labus (210 BC), where the light infantry javelinmen are collectively referred to as *akontistai*.[30]

Onasander groups the *akontistai* along with slingers (*sphendónitai*) and archers (*toxotai*) under the *psiloi*.[31] He uses the term *akontistai* only once, when he discusses where a general would deploy javelinmen in relation to the heavy infantry. As with archers and slingers, Onasander advocates that javelineers should be deployed in front of the heavy infantry:

> '[F]or if placed in the rear they will do more damage to their own army than to the enemy, and if in among the heavy-armed, their peculiar skill will be ineffectual because they will be unable to take a step backwards in throwing their javelins or to charge forward and cast them.'[32]

Plutarch, one of our later sources on the period under discussion, also uses *akontistai* to refer to javelinmen in the armies of fifth-century BC Athens (*Nic.*, 24.4), Carthaginian forces in the second century BC (*Marc.*, 29.3) and the late Republican forces of Mark Antony in the first century BC (*Ant.*, 41.4, 42.1), further emphasizing that the term was not restricted by time or ethnicity.

The term *akontistai* was still used to describe javelinmen in the Imperial Roman armies by authors such as Arrian and Josephus, and in the Byzantine military manuals of the medieval era, as mentioned previously.[33]

4.1 Arms and Armour of the Akontistai

4.1.1 Javelin and the Throwing Strap

The *akōntion* (javelin) was primarily a missile weapon as opposed to a throwing spear, such as the dual-purpose *longchē* which could be thrown as a missile as well as used in hand-to-hand combat (**see plate 22**). The *akōntion* is generally associated with a throwing strap called an *ankylē* (ἀγκύλη) in Greek and *amentum* in Latin, which was a leather thong attached to the shaft of the javelin that assisted in the casting of the weapon (**see plate 23**). It is attested in the literary sources as well as in iconographical evidence. The Alexander Mosaic serves as a beautiful example of a javelin with an *ankylē* throwing strap attached. In the foreground of the mosaic lies a javelin which missed its mark, or perhaps one simply discarded in the chaos of battle, with its attached *ankylē* clearly visible (bottom centre of the mosaic).[34]

Because it was primarily a missile weapon, the shaft of the javelin was fairly thin and its head small to make it as aerodynamic as possible. The javelins used by the Roman *velites* were, according to Polybius, 'about a finger's breadth in thickness', which is just under 2cm (19.3mm).[35] The thickness of the shaft played a role not only in how aerodynamic a javelin was, but in how many javelins could be carried. *Akontistai* carried their spare javelins in their non-throwing hand. One would imagine that if the shafts of the javelins were thicker than 'a finger's breadth in thickness', it would severely limit the amount of them a man could carry. While there is very little information on how many javelins the *akontistai* carried, there is one reference to Roman light infantry carrying seven javelins.[36] Javelin shafts thus had to be fairly thin for a soldier to carry at least six in one hand, the seventh being in his throwing hand. Both Xenophon and Theophrastus state that javelin shafts were made from cornel wood.[37] Cornel is a hard wood with a high density. This made the shaft heavier despite it being only 'a finger's breadth in thickness', which allowed the javelin to strike its target with more impact and kinetic energy.

Yet it must be remembered that there were a great many different types of javelin throughout our period of discussion and across the various cultures. For example, the Iberian *soliferreum*, prominent across the Spanish peninsula, was an all-iron javelin which was 1cm in diameter and measured between 1.6 and 2 metres, with a barbed head.[38] The javelins of a Samnite warrior portrayed on the Nola Tomb frieze are shown as taller than the warrior and about 'a finger's breadth in thickness'.[39] Some Germanic javelins, according to Tacitus, were entirely made of wood, with a sharp, fire-hardened point. Roman *grosphos* used by the *velites* had a wooden shaft of 2 cubits (96cm) with an iron head a span (22.86cm) in length. Polybius describes the javelin as follows:

> 'The *grosphos* of the *velites* has a wooden shaft of about two cubits, and about a finger's breadth in thickness; its head is a span long, hammered fine, and sharpened to such an extent that it becomes bent the first time it strikes, and cannot be used by the enemy to hurl back; otherwise the weapon would be available for both sides alike.'[40]

4.1.2 Protective Gear

Akontistai, like most light infantry, as we have seen, were generally recruited from the lower classes of society. The financial restriction that their status imposed upon them grouped them among the *psiloi*, which meant that they were generally unarmoured; as stated by Aclepiodotus, the *psiloi* 'uses the lightest equipment because it shoots from a distance, and is provided with neither greaves nor cuirasses'.[41] There are several depictions of light infantry with cloaks or animal pelts draped from their arms to serve as protection to catch incoming missiles or defend against other forms of attack. This was not unlike the curtains which were sometimes attached to the hoplites' *aspis* shields. A prime example is from an Athenian-manufactured drinking cup, now in the British Museum, showing a naked youth with a javelin in his right hand and a cloak wrapped around his left arm.[42] There is also the kylix of a javelin-armed youth with a leopard skin draped over his left arm. Other than a broad-rimmed hat, he is completely naked.[43] The use of cloaks draped from the left arm is also attested in literature. As mentioned, *akontistai* was a broad and general term for javelinmen, and those who were more well-to-do could afford better equipment. We have already seen that Polybius used *akontistai* as one of several terms to refer to the *velites*, the Roman Republican citizen infantry (see also Chapter 5: Velites/Grosphomachoi). Wearing helmets and carrying both shields and swords, the *velites* would be much better equipped than, for example, an Athenian levied *akontistai* or a Spartan javelin-armed helot from the fifth century BC. The *peltasts*, generally speaking, were also javelinmen but better trained and equipped than the standard *akontistai.*

Classical Peltast

Peltasts or *peltastes* (πελταστής), sometimes also referred to as *peltastai* (πελτασταί), were light infantry who originated among the wild Thracian tribes of northern Greece. They were primarily armed with javelins and carried into combat a small, often crescent-shape wicker shield called a *peltē* (πέλτη) **(see plate 24)**. It is from this shield that *peltasts* derived their name. Because of their homeland's economic limitations and its mountainous and difficult terrain, Thracians were natural light infantry and the *peltasts* formed the backbone of Thracian armies. Their prominence in Greek warfare grew during the Peloponnesian War, due to their successful use against hoplites. They first served as mercenaries hired from their place of origin, but the Greek city-states, quick to recognize the effectiveness and potential of the particular style of fighting, started training and rearming their own light troops as *peltasts.*

While Graeco-Roman warfare continued to be dominated by the heavy infantry, the predominance of difficult rough terrain, the necessity of forced marches, the need for reconnaissance, range warfare, protection of the marching column, raiding, and attacking and defending fortified positions and cities, highlighted the importance and need for light troops. It was the poor who generally filled the roles of light infantry. This was very much because their social and economic status did not afford the training or quality equipment needed for heavy infantry or cavalry roles, nor did their loftier peers of the higher classes afford them these. What separated the *peltast* from other javelineers was that the *peltasts* were specialists in their style of combat. They were better trained and equipped than the poor, lower-class irregular *akontistai*. In the words of Thucydides, *peltasts* were 'prepared light-armed troops'. Being better equipped and trained allowed these troops, under the right circumstances, to engage in melees, making them a very versatile unit in combat. Iconographical evidence shows *peltasts* sometimes armed with swords as secondary weapons and using their spears or javelins in a melee. The *peltast* thus fell somewhere between the general light infantry and the heavy infantry. The military manuals of Asclepiodotus, Aelian and Arrian all make a clear distinction between light troops and *peltasts*, who were in a class of their own, separating them from the other light infantry.[44] Asclepiodotus writes:

> 'The infantry is divided into the corps of hoplites, the corps of *peltasts*, and the corps of so-called light infantry [*psiloi*]. Now the corps of hoplites, since it fights at close quarters, uses very heavy equipment – for the men are protected by shields of the largest size, cuirasses, and greaves – and long spears of the type which will here be called "Macedonian". The corps of the light infantry on the contrary uses the lightest equipment because it shoots from a distance, and is provided with neither greaves nor cuirasses, but javelins and slings, and in general with missiles which we call "long-distance missiles". The corps of the *peltasts* stands in a sense between these two, for the *peltē* is a kind of small, light shield, and their spears are much shorter than those of the hoplites.'[45]

While the writings of Asclepiodotus, Aelian, and Arrian date between first century BC and the second century AD, the distinctions between *peltasts* and the other *psiloi* are clear and equally relevant to earlier periods.

As mentioned, the *peltasts* came into their own right during the Peloponnesian War (431–405 BC), the term *peltast* appearing several times in the writing of Thucydides. The reason for the increased reliance on *peltasts* during the conflict between Athens and Sparta came down to the new demands created by the unprecedented scale of the war, both in chronological and geographical terms. From 425 BC, the war shifted tremendously from how Sparta and Athens originally anticipated. The conflict moved further and further away from the belligerents' home ground, being fought in the lesser-known peripheral regions of Greece and even further afield. Because the conflict was fought far from Sparta and Athens and spread into the lands of their respective allies, there was a greater reliance on these allies to fill the ranks around an ever-smaller core of Spartan and Athenian hoplite contingents. Many of the allies of Athens and Sparta lacked the means to field large contingents of heavy infantry, so the troops they supplied were of the *peltast* type. *Peltasts* were far cheaper, being paid a mere three *obols* a day, meaning more soldiers could be fielded, and those who served as mercenaries could fight further from home and stay in the field much longer than citizen soldiers.[46] Even the landless poor could probably do better as paid *psiloi* than making do with what they could scratch together back home.

Peltasts first appear in the pages of Thucydides' *The Peloponnesian War* when Sitalces, the king of the Thracian Odrysians, due to his newly formed alliance with Athens, promised to send a force of Thracian cavalry and *peltasts* to aid Athens in its war.[47] We next see *peltasts* in the summer of 429 BC, when an Athenian force clashed with a force of Boeotians and Chalcidians, where *peltasts* were fielded on both sides.[48] Cleon's force with which he defeated the Spartans at Sphacteria (425 BC) contained a large contingent of *peltasts*.[49] When the Athenians invaded Boeotia in 424 BC, the defending army which marched out to meet them numbered 7,000 hoplites, 1,000 cavalry, 10,000 light troops and 500 *peltasts*.[50] This section is of particular interest because Thucydides clearly distinguishes between the *peltasts* and the rest of the light troops, not unlike the later military manuals of Asclepiodotus, Aelian and Arrian.

Peltasts were extensively used by the exceptional Spartan general Brasidas in his northern campaigns between 424 and 422 BC. The *peltasts* played a pivotal role in his capture of the city of Torone through a daring surprise attack.[51] When Brasidas sent troops to help defend the cities of Mende and Scione against an Athenian attack, they consisted of 500 hoplites and 300 *peltasts*.[52] The expected Athenian counterattack comprised 1,000 hoplites, 600 archers, 1,000 Thracian mercenaries (in all likelihood *peltasts*) and an unnumbered contingent of allied *peltasts*.[53] At the Second Battle of Amphipolis (422 BC), Brasidas' 2,000 hoplites and 300 cavalry were supported by a significant force of *peltasts*, which consisted of 1,500 Thracian *peltasts*, 1,000 *peltasts* from Myrcinus, Amphipolis and Chalcidia, and an additional unnumbered force of *peltasts* from Edonia.[54] Brasidas managed to gain a victory over the superior (in quality, not number) Athenian army of Cleon by means of a sudden sally that caught the Athenians off guard. The *peltasts* and cavalry killed many of the fleeing Athenians, Thucydides (5.10.9) stating that 'Cleon himself, who from the start had never intended to stand his ground, fled immediately and was overtaken and killed by a *peltast* from Myrcinus.'

In preparation for the Sicilian expedition, Demosthenes hired a Thracian mercenary contingent of *peltasts* 'from the sword-bearing [*machairophoroi*] tribe known as the Dians/Dii'.[55] However, they arrived too late to join the expedition and were sent back to Thrace. Their very name, 'sword-bearing', perhaps supports the idea that *peltasts* were not restricted to being range units but were willing to engage in melees when the opportunity presented itself. While the Thracian Dians did not accompany Demosthenes to Sicily, they did attack Athens' enemy, the Boeotian city of Mycalessus, and butchered the entire population – men, women and children. The massacre was described by Thucydides (7.29.5) as 'unsurpassed in magnitude, and unapproached by any in suddenness and in horror'. When the Thebans heard of the disaster, they sent a force of horsemen to attack the Thracians. These Thracians, however, gave a good account of themselves against the mounted troops, demonstrating their skill and professionalism and showing that trained light troops could even hold their own against cavalry. The fact that they were able to form disciplined ranks places further emphasis on their level of training:

> '[During] the rest of the retreat the Thracians made a very respectable defence against the Theban horse, by which they were first attacked, dashing out and closing their ranks according to the tactics of their country, and lost only a few men in that part of the affair.'[56]

The Peloponnesian War highlighted among the Greeks the need for light infantry, and because of the growth of their importance, a greater emphasis was placed upon the specialization and training of these troops. The *peltast* was the natural step taken to making light troops – *akontistai* in particular – more professional and able to fill the gaps left by the limitations of the citizen hoplite.

Peltasts also appear several times in Xenophon's writings. However, in contrast to Thucydides, who uses *peltast* to refer to a specific unit type, Xenophon often employs the term to collectively refer to light infantry. This is likely because of the growing reliance on and importance of light troops and their continued professionalization. The need for light troops could simply no longer be denied, even by conservative Sparta. So when Xenophon uses the term *peltast*, it is to reference a more professional and better-trained and equipped light infantry rather than their particular armament or a specific unit type. For instance, in 403 BC, the Athenian light troops which resisted the Thirty Tyrants and fought against Pausanias at Piraeus are collectively called '*peltastai*', and in the very next line referred to as '*psiloi*' who made use of all manner of ranged weapons, 'throwing javelins, hurling stones, shooting arrows, and discharging slings'.[57] This also occurs throughout the *Anabasis*.[58] There are, however, certain sections where Xenophon does distinguish between bowmen, slingers and true javelin-armed *peltasts*:

> 'Xenophon passed word to all the peltasts to advance with hand on the thong, so that they could discharge their javelins when the signal should

> be given, to the bowmen to have their arrows upon the string, ready to shoot upon the signal, and to the slingers to have their bags full of stones.'[59]

There are two particularly interesting depictions of Greek *peltasts*. The first is from the Altıkulaç Sarcophagus, which dates between 400 and 375 BC. The sarcophagus depicts an Anatolian dynastic cavalryman riding down a Greek light infantryman. To the rear of the dynast is another light infantryman. Scholars have identified him as a Greek mercenary in the dynast's employ.[60] In his left hand he bears a small round shield, which can be identified as a *peltē* (**see plate 25**). He also holds two light javelins in his shield hand. In his right hand he holds a drawn curved sabre of the *machaira* or *kopis* type (**see plate 26**). This goes a long way to demonstrate the willingness and ability of the *peltast* to engage in a melee. He is clothed in a long-sleeved tunic, but other than that he wears no armour. The Greek soldier who has been ridden down by the dynast can also be identified as a *peltast*. Similarly to the mercenary, the Greek soldier holds a small *peltē* in his left hand. Likewise, he is unarmoured. He has no javelins; they might have already been spent in the battle or discarded in his attempt to flee the horsemen. Down but not out, his right hand reaches for a sword hanging from his left hip, further solidifying the *peltast*'s melee fighting potential.

The second example comes from the Tomb of Payava (*circa* 360 BC), which displays a battle scene with three Greek *peltasts* fighting among cavalry and other soldiers. They wear *pilos* helmets, two of which are crested, but otherwise they wear no armour. The two *peltasts* with crested helmets are clothed in the *exomis* tunic, with the third being completely naked. All three carry small round rimless *peltai* shields. Their right arms are held high above their heads in what appears to be an overarm thrusting method. Unfortunately, their weapons are eroded away, but it is possible they were armed with thrusting spears, javelin ready to be thrown or even swords. Some scholars have interpreted them as Iphicratic *peltasts*, but their lack of any form of armour, boots and, more particularly, the long Iphicratic pike suggest that they were not *peltasts* of the Iphicratic reform (see below: The Reforms of Iphicrates).

As with all light infantry, the *peltast* suffered the same prejudice. Plutarch's *Comparison of Lysander and Sulla* (4.2–3) serves as a prime example:

> 'For he [Lysander] received his death wound, not as Cleombrotus did, at Leuctra, standing firm against the enemy's onsets, nor as Cyrus did, or Epaminondas, rallying his men and assuring the victory to them; these all died the death of kings and generals. But Lysander threw away his life in gloriously, like a common *peltast* or skirmisher.'

The passage makes it clear that because of their fighting style, *peltasts* and light troops in general were unable to obtain a glorious and noble death like the heavy infantry heroes who died in pitched battle. Their effectiveness in war, however, could not be denied, the increased reliance on *peltasts* in the Peloponnesian War and after being testament to this.

4.2 Arms and Armour of the Peltasts

4.2.1 The Peltē

The *peltē* was often crescent-shaped, but oval and circular shields are also depicted (**see plate 25**).[61] Shields tended to be rimless and have a wicker or wood base, being covered with animal skin or felt. Aristotle mentions that they were covered in goatskin rather than oxhide or bronze.[62] This is most likely because goatskin was more readily available and far lighter than oxhide. Bronze and even gold are also attested, but these would be rarer and restricted to highly specialized, wealthier light troops. The Cretan mercenary archers in the Greek army of Xenophon's *Anabasis* are, for instance, armed with bronze *peltai*.[63] Bronze-covered *peltai* become far more common during the Hellenistic era, as attested in the military manuals of Asclepiodotus and Aelian (**see plate 25**).[64] There is visual evidence of *peltai* shields, the majority appearing on Greek pottery. The shield face was often decorated with geometrical patterns, crude faces with large eyes and even animals. The size of the *peltē* was not standardized, with smaller and larger versions attested in art. The majority, however, are generally shown to be approximately 50cm in length. The shield was attached to the forearm by means of leather straps and had an off-centre handgrip. Its shape, size and light material allowed for the shield to be wielded with ease and contributed to the mobility of the *peltast*.

4.2.2 Thracian Cloak

Most depictions of Thracian *peltasts* show them wearing a distinctive thick cloak called a *zeira* (**see plate 24**). These cloaks were long, hanging from the shoulders all the way to the ankles, with intricate geometrical patterns on them. From the iconographical evidence, they were clearly worn during combat. The protection such a cloak provided must not be underestimated. The loose hanging garment would significantly reduce the kinetic energy that a light javelin, arrow and in particular a light lead slingshot could impact upon its intended target. Again, this was not all that different from shield-curtains or the curtain hung from triremes to protect the rowers from missiles. The cloak would especially be effective when the *peltast* turned his back to flee from an advancing enemy. At the Battle of Telamon (225 BC), Polybius gives an interesting account of the protection cloaks can afford a warrior. We are told that the majority of Gallic soldiers were fairly well protected from the javelins thrown by the *velites* because of their cloaks and other garments. However, the *Gaesatae*, who fought naked, suffered tremendously because of their lack of clothing:

> 'The Insubres and Boii wore their trousers and cloaks, but the Gaesatae had discarded these garments owing to their proud confidence in themselves, and stood naked, with nothing but their arms, in front of the whole army, thinking that thus they would be more efficient, as some of the ground was overgrown with brambles which would catch in their clothes and impede the use of their weapons ... But when the javelineers [*akontistai*] advanced, as is their usage, from the ranks of the Roman legions and began to hurl their javelins in well-aimed volleys, the Celts in the rear ranks indeed were well protected by their trousers and cloaks, but it fell out far otherwise than they had expected with the naked men in front, and they found themselves in a very difficult and helpless predicament. For the Gaulish shield does not cover the whole body; so that their nakedness was a disadvantage, and the bigger they were the better chance had the missiles of going home.'[65]

While the account does not specifically refer to the Thracian cloak, it goes to some length to show how effective such a cloak could be. It does not seem that the cloak was adopted by non-Thracian *peltasts*.

4.2.3 Javelins/Throwing Spears

Most scholars seem to accept that Thracian *peltasts* were armed with only two javelins. This is primarily based on the iconographical evidence, which always shows Thracian *peltasts* armed with one or two javelins, but never more. This might just be artistic licence, and it is possible that more than two javelins were carried. The size – both length and thickness – of Thracian javelins portrayed in art, however, suggests that it would be difficult to carry more than two, as it would restrict the mobility of the *peltast*. The length and thickness of the javelin is also more akin to those of a spear, like that of the dual-purpose *longchē*, meaning it could be used as both a missile and a melee weapon (**see plate 24**). As mentioned, there are several images of Thracian *peltasts* using their throwing spears in hand-to-hand fighting.[66] When they did use their javelin in a melee, their shield would at least give the *peltasts* some staying power. The *peltasts* of Greek origin seem to have retained the shorter, lighter *akōntion* javelin, like the *peltast* depicted on the Altıkulaç Sarcophagus, which was far more suited as a missile weapon. These lighter javelins were undoubtedly thrown with the *ankylē* throwing strap, as highlighted in the *Anabasis*: 'Xenophon passed word to all the *peltasts* to advance with hand on the thong, so that they could discharge their javelins when the signal should be given.'[67]

The javelin was the most fitting weapon for the *peltast*. While better equipped and trained, and more disciplined, than the basic *psiloi*, they were still too lightly armoured to fight heavy infantry head on, and were thus primarily a range unit. Archery was a highly disciplined skill that took years to master. As discussed above, the bow and arrow was to a large extent restricted to mercenaries who came from a culture of archery, such as those from Crete and Scythia. The deployment of citizen archers, such as those fielded by Athens, were thus at best a rarity. The construction of the bow was time-consuming and required specialized expertise, and the bow itself was sensitive to the climate. The sling, while very inexpensive, also required practise and skill to use, and slingers were generally hired from the lowest strata of society, often from pastoral peoples whose daily life required the use of a sling to protect their flocks. The javelin, meanwhile, was far cheaper to construct than the bow and much easier to master. It was very much part of Greek and Roman culture, serving as the primary hunting tool, and even formed part of their athletics training and competitions. Indeed, most men in the Graeco-Roman world would have had some experience throwing javelins.

4.2.4 Swords

The *peltast* was not restricted to fighting with missiles, but could engage in a melee if the opportunity presented itself. As a secondary weapon, the *peltast* could be armed with a knife or sword. We have already come across the Thracian Dians/Dii, whom Thucydides referred to as 'sword-bearing [*machairophoroi*]'. We have also seen that both *peltasts* on the Altıkulaç Sarcophagus were armed with swords. Greek swords are generally of two types: the straight-bladed cut-and-thrust *xiphos* and the forward-curved *makhaira* (also called a *kopis*) ideally suited for cutting (**see plate 26**). While the sword of the Greek *peltast* to the right on the sarcophagus is impossible to make out, the mercenary Greek *peltast* fighting alongside the Anatolian dynast has a drawn sword of the *machaira/kopis* type. The *machaira* sword had a heavy forward-curved single edge cutting blade, and is widely attested in pottery and other imagery in the hands of both Greeks and Persians. Its popularity spread, and it was adopted by the tribes on the Italian peninsula and the Iberian tribesmen in Spain.[68] From Xenophon's description of cavalry swords, it is possible that the Greeks adopted the sword from the Achaemenid Persians:

> 'For harming the enemy we recommend the sabre [*makhaira*] rather than the sword [*xiphos*], because, owing to his lofty position, the rider will find the cut with the Persian sabre [*kopis*] more efficacious than the thrust with the sword.'[69]

The straight-bladed *xiphos* is very well attested in iconography. Numerous Greek hoplites are portrayed with the *xiphos* sword in sixth- and fifth-century BC pottery. It also appears frequently in Hellenistic imagery, and Alexander the Great and his *hetairoi* bodyguard are often shown armed with the straight-bladed *xiphos*.[70] According to Diodorus, the later Iphicratic *peltasts* were armed with long *xiphe* swords. The *xiphos* was thus a very likely option for a secondary weapon for these more professional *peltasts*.

The Reforms of Iphicrates

The Athenian general Iphicrates (418–353 BC), despite his humble beginnings as the son of a shoemaker, became one of the most renowned military figures of his time. Cornelius Nepos stated:

> 'Iphicrates of Athens has become renowned, not so much for the greatness of his exploits, as for his knowledge of military tactics; for he was such a leader, that he was not only comparable to the first commanders of his own time, but no one even of the older generals could be set above him.'[71]

His renown was built upon him being a commander of light troops, his reputation being sealed in the Corinthian War (395–387 BC) at the Battle of Lechaeum (391 BC), where he successfully used his mercenary *peltasts* to annihilate an entire *mora* (battalion of 600 men) of Sparta's best hoplites. Having been surrounded by the *peltasts*, and without light troops of their own, the Spartan hoplites had no answer for the hit-and-run tactics of the *peltasts*, whom Xenophon describes as the 'nimblest of all troops'.

Light troops could simply no longer be ignored, and armies could not now take the field without their support. The stage had been set for combined-arms militaries, something clearly recognized by Iphicrates:

> '[A]s Iphicrates analysed the matter, the light-armed troops are like the hands, the cavalry like the feet, the line of men-at-arms itself like chest and breastplate, and the general like the head.'[72]

Iphicrates also recognized that the phalanx of hoplites, which dominated Greek warfare, had become redundant and archaic, and desperately needed to change due to the permanence of light troops. The general thus sought to bridge the gap between heavy and light infantry by reforming the armament of the Classical hoplite. Cornelius Nepos gives a detailed description of these reforms:

> 'He altered the arms of the infantry; for whereas, before he became a commander, they used very large shields, short spears, and small swords, he, on the contrary, introduced the *pelta* instead of the *parma* (from which the infantry were afterwards called *peltastae*), that they might be more active in movements and encounters; he doubled the length of the spear, and made the swords also longer. He likewise changed the character of their cuirasses, and gave them linen ones instead of those of bronze; a change by which he rendered the soldiers more active; for, diminishing the weight, he provided what would equally protect the body, and be light.'[73]

Diodorus Siculus also highlights Iphicrates' reforms:

> 'The Greeks were using shields [*aspidai*][74] which were large and consequently difficult to handle; these he discarded and made small oval ones of moderate size, thus successfully achieving both objects, to furnish the body with adequate cover and to enable the user of the small shield, on account of its lightness, to be completely free in his movements. After a trial of the new shield its easy manipulation secured its adoption, and the infantry who had formerly been called "hoplites" because of their heavy shield, then had their name changed to "*peltasts*" from the light *peltē* they carried. As regards spear and sword, he made changes in the contrary direction: namely, he increased the length of the spears by half, and made the swords almost twice as long. The actual use of these arms confirmed the initial test and from the success of the experiment won great fame for the inventive genius of the general. He made soldiers' boots that were easy to untie and light and they continue to this day to be called "*iphicratids*" after him.'[75]

The purpose of the reforms is clear: by reducing the size and weight of the shield, changing the cuirass from bronze to lighter and more flexible linen, and exchanging their greaves for flexible high boots, their movement was no longer so restricted, making the hoplite much more agile and nimble. The protective gear thus reflected that of light infantry. Their main focus, however, remained melee combat. The length of their spears and swords were increased, so while they lost some defensive ability with the reduction in the size of the shield and the exchange of a metal cuirass for a linen one, they gained a greater range advantage with their weaponry.

Iphicrates' innovations were not only based on making the hoplite more agile – the economic implications of the reforms must not be overlooked. The hoplite's equipment was very expensive, being restricted to the upper-middle classes of the more well-to-do Greek city-states. I have already discussed how many of the allies of Sparta and Athens and other poorer Greek states simply did not have the means to field large hoplite armies and relied primarily upon *peltasts* for the backbone of their military forces. But while the Classical hoplite phalanx was slowly becoming obsolete, the need for heavy infantry was not. The heavy infantry continued to dominate the major pitched battles of the ancient world. Although the importance of and need for light troops grew, there was still a requirement to field melee troops. Reducing the hoplite's shield size and substituting the metal cuirass with a linen corslet also reduced the cost of equipment of fielding heavy infantry. Iphicrates was primarily a commander of light troops, and these were mainly mercenaries. While there was the odd man who might have joined a mercenary expedition for lust of adventure, the vast majority of mercenaries were forced to take on the role of a soldier of fortune out of financial desperation. These men would join as *peltasts* as they could not afford the expensive hoplite gear. The reforms were thus just as much a financial means of bridging the gap between light and heavy infantry.

When one looks at Iphicrates' *peltasts*, one cannot help but picture the Hellenistic Macedonian phalanx of Philip II and Alexander the Great, with men bearing their long *sarissa* pikes and their *peltē* shields strapped to their left arms. Many scholars now believe that Iphicrates' reforms were the foundation for the Macedonian-style phalanx which dominated the Hellenistic era.[76] The new Iphicratic *peltast* clearly now focused on melee fighting, in contrast to those prior to the reforms who focused on range combat.

4.3 Arms and Armour of the Iphicratic Peltast

4.3.1 The Iphicratic Peltē

According to Nepos and Diodorus, the *aspis* (or *parma* in Latin) was replaced by the *peltē/pelta*. Nepos describes the *parma/aspis* as 'very large shields', while Diodorus says it was 'large and consequently difficult to handle'. It was round, approximately 80cm in diameter, with a wide rim. It had a wooden core with a thin covering of bronze, and weighed 7.5kg.[77] The new *peltē* was obviously much smaller and lighter, Diodorus describing it as a 'small oval' and 'of moderate size'. The size of the Iphicratic *peltē* is unknown, but Asclepiodotus states that the Hellenistic *peltē* was 8 palms (64cm) in diameter (**see plate 25**). These *peltai* could provide their wearer with adequate cover, yet were light enough to allow the Iphicratic *peltasts* to move lightly with minimal restriction, or in the words of Nepos, to 'be more active in movements and encounters'. It is most likely that as with the Thracian *peltē*, the Iphicratic *peltē* had a simple wood or wicker base covered in hide, as opposed to bronze. The financial restrictions of fourth-century BC mercenaries would probably not allow for a bronze covering like the *peltē* of the following centuries. Since the focus shifted to more melee-type light troops, it is possible that the core of the shield was wood rather than wicker, making it more durable for hand-to-hand combat.

4.3.2 Iphicratic Spear

Both Nepos and Diodorus emphasize the greater length of the Iphicratic *peltast*'s spear in comparison to that *doru* of the hoplite. The exact length of hoplite *doru* is impossible to determine, but it lay somewhere between 170cm and 250cm.[78] Because of the size of the *aspis* shield, the entire left arm was needed to bear the shield's weight, meaning the *doru* was wielded one-handed in the free right hand. Nepos states that Iphicrates doubled the length of the spear, meaning it would be approximately 340–500cm. This is far more akin to a pike than a spear. Diodorus, however, writes that Iphircrates only increased the length of the spear by half, making it approximately 255–375cm. Regardless, the sheer length of the spear, whether it was 255cm or 500cm, clearly made it a melee weapon, not suitable for throwing. The longer spear thus gave the Iphicratic *peltast* a reach advantage which allowed him to keep his enemy at a distance. Again, it is hard to deny the similarities between the Iphicratic *peltast* and the later Macedonian *phalangite* of Philip II and Alexander.

4.3.3 Sword

As with the spear, both accounts of the Iphicratic reforms state that the length of the *peltast*'s sword was increased. While Nepos only mentions that the sword was also lengthened, Diodorus states that the sword length was doubled. The word used for sword in Diodorus' text is *xiphos* (ξίφος), a term usually associated with a straight-bladed, double-edged Greek sword. *Xiphe* swords were far from standardized, their blades coming in different shapes and sizes. Some blades were completely straight and narrow, while others swelled towards the midsection before tapering towards the point, giving it a leaf shape. They ranged from as short as 30cm to almost 70cm in length. Iconographical depictions of hoplites show them carrying *xiphe* swords of all lengths and shapes, which suggests personal preference. The shorter versions were often preferred by hoplites because of the special restrictions created by hoplite combat. The Spartan hoplites, for instance, preferred a much shorter *xiphos* than their Athenian counterparts. Plutarch (Lyc., 19.2) recounts the tale of King Agis III of Sparta being mocked by an Athenian over the size of their swords, stating that the Spartan sword was so small 'that jugglers on the stage easily swallowed them'. Agis retorted in true Spartan fashion, 'yet we certainly reach our enemies with these daggers'. I am of the opinion that the lengthening of the sword is simply a case of replacing shorter swords like those used by the Spartans with longer versions, rather than lengthening the already longer versions. Greek swords, especially of the *xiphos* type, don't exceed 70cm, there being no archaeological or iconographical evidence of any longer swords. There is also the possibility that Iphicrates simply replaced knives or daggers with swords. As we have seen, Greek military terminology was far from standardized; another Greek term for sword, *machaira*, which was generally used to refer to curved sabres such as the *kopis*, was also used for a variety of knives, the Roman *gladius* and the straight-bladed Gallic longsword. It must be noted that I have yet to find a use of the term *xiphos* to denote anything other than a sword. Nevertheless, what we can take from the lengthening of the sword is the greater emphasis placed on the melee capabilities of these reformed *peltasts*.

4.3.4 Linen armour

There is no mention of body armour in the account of Diodorus, but Nepos does say that the Iphicratic *peltasts* substituted the bronze cuirass with a linen one (**see plate 12**). This change, Nepos says, 'rendered the soldiers more active'. A corslet of linen was lighter, cheaper and imposed far less restriction of movement upon its wearer than one made of bronze. As discussed above, the *lorica lintea* in Latin (*linothōrax* in Greek) was a popular form of light armour throughout the Classical and Hellenistic era and many heavy infantry soldiers also opted for this type of body protection.

4.3.5 Iphicratids

Nepos makes no mention of *Iphicratids* boots, but Diodorus comments upon the newly adopted boots which 'were easy to untie and light'. The very fact that these boots were named after Iphicrates and still referred to as *Iphicratids* in the first century BC makes it clear that they were highly effective and fairly popular. One would imagine that these boots replaced the bronze greaves or *knēmides* (κνημῖδες), which would once again aid in mobility and still provide protection to the vulnerable shins. There are several images of Thracian *peltasts* wearing boots that pull up to just under the knee.

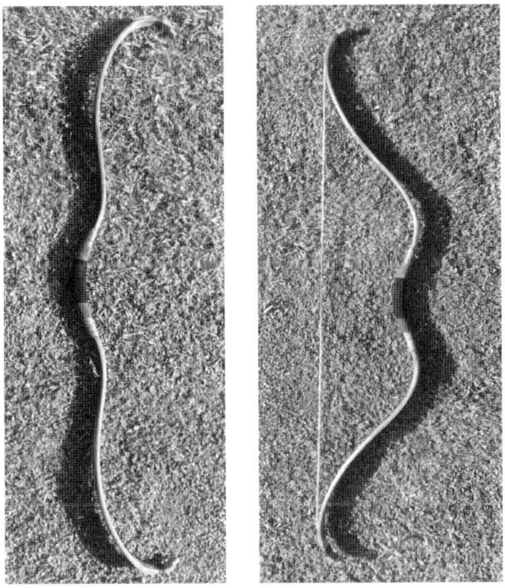

1. Bow unstrung (left) and strung (right). (*Author's collection*)

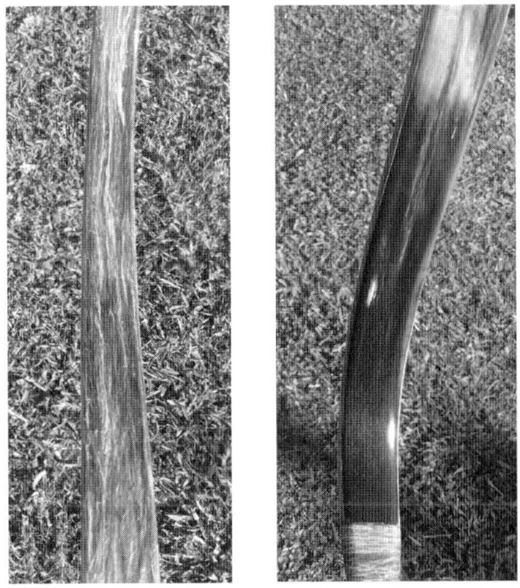

2. Sinew-covered 'back' (left) and horn-layered 'belly' (right) of a bow. (*Author's collection*)

3. Bow in full draw. (*Author's collection*)

4. Scythian archer depicted on Greek kylix (550–520 BC). (*Wikimedia Commons/ Public domain*)

5. Elamite/Persian horn bow. (*Author's collection*)

6. Persian Immortals on the Frieze of Archers. (*Pergamon Museum, Berlin via Wikimedia Commons /CC BY 2.0*)

7. Imperial guardsmen portrayed on bas-relief of Persepolis. Note the 'Scythian' bow in the *gorytos* and the 'Elamite/Persian' bow slung across the shoulder. (*Public domain*)

8. Mediterranean draw. (*Author's collection*)

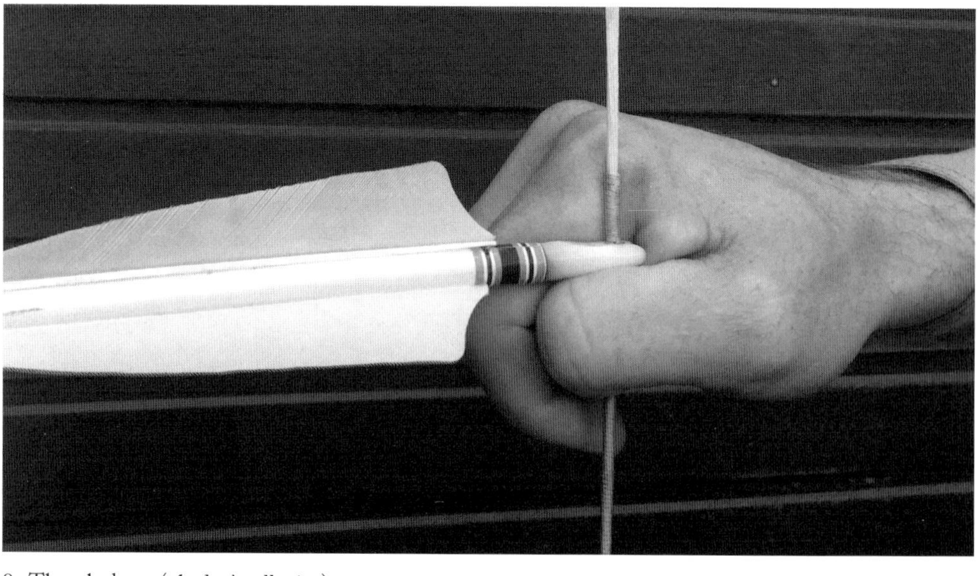

9. Thumb draw. (*Author's collection*)

10. Odysseus slaying the suitors. Note Odysseus's use of a variation of the thumb draw. (*Public domain*)

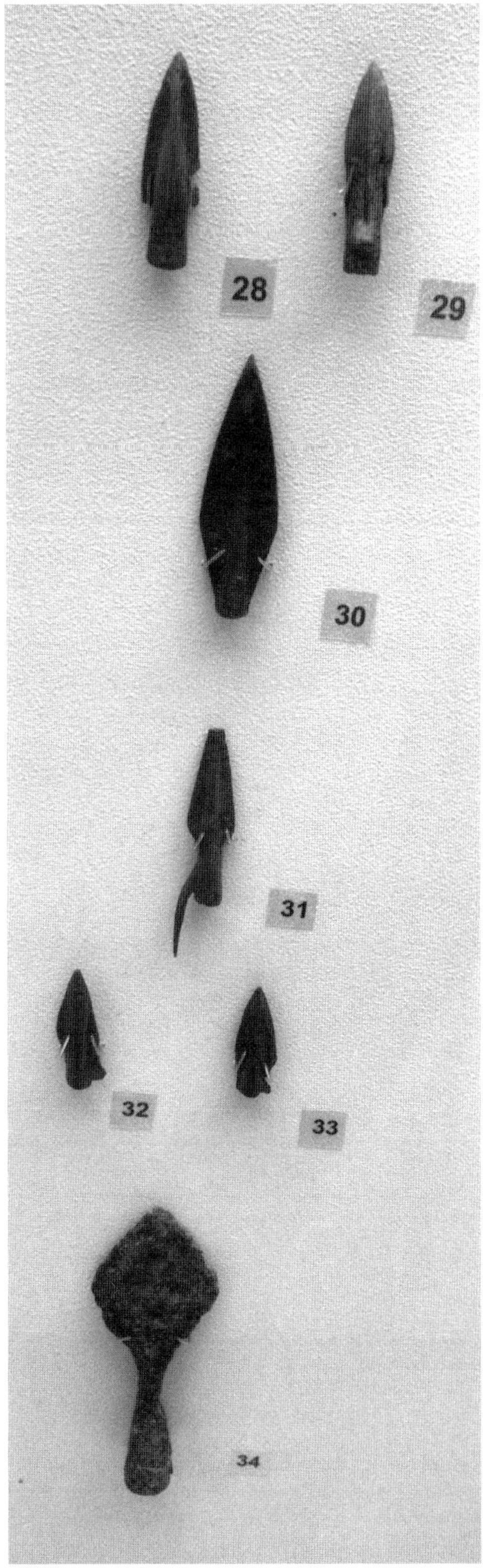

11. A variety of different arrowheads from the Graeco-Roman period. (*Author's collection*)

12. Reconstruction of the *linothōrax* body armour. (*Author's collection*)

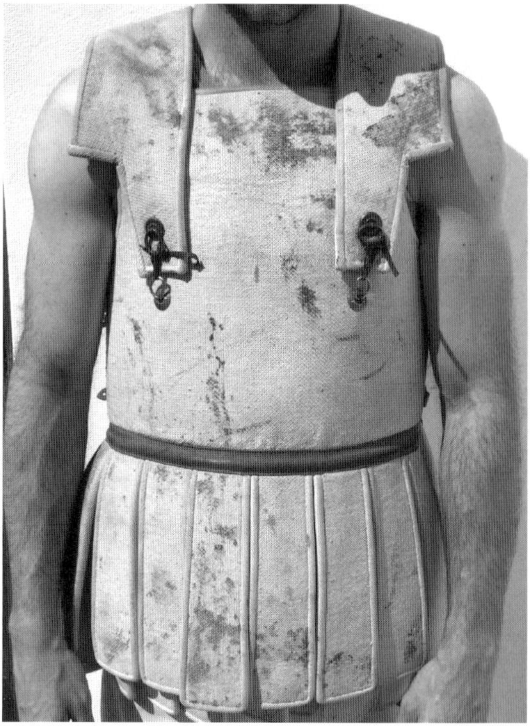

13. *Peltē* shield and bow combination. (*Author's collection*)

14. While images of slingers are rare, the hero Heracles is portrayed slaying the Stymphalan birds with a sling as opposed to being traditionally armed with a bow. (*Public domain*)

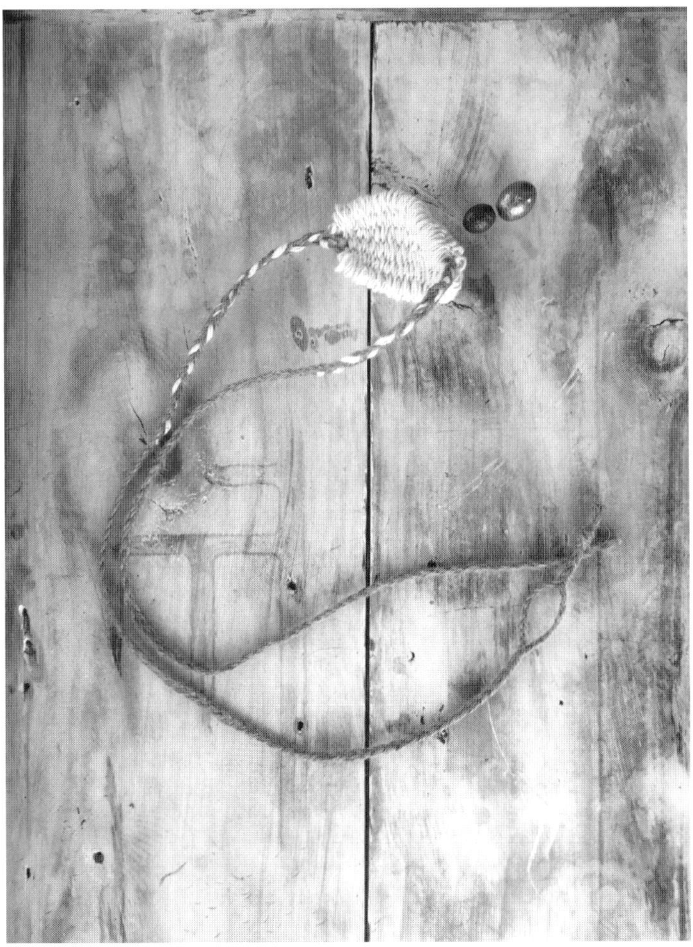

15. Sling of 'triple' construction made of hemp and wool. (*Author's collection*)

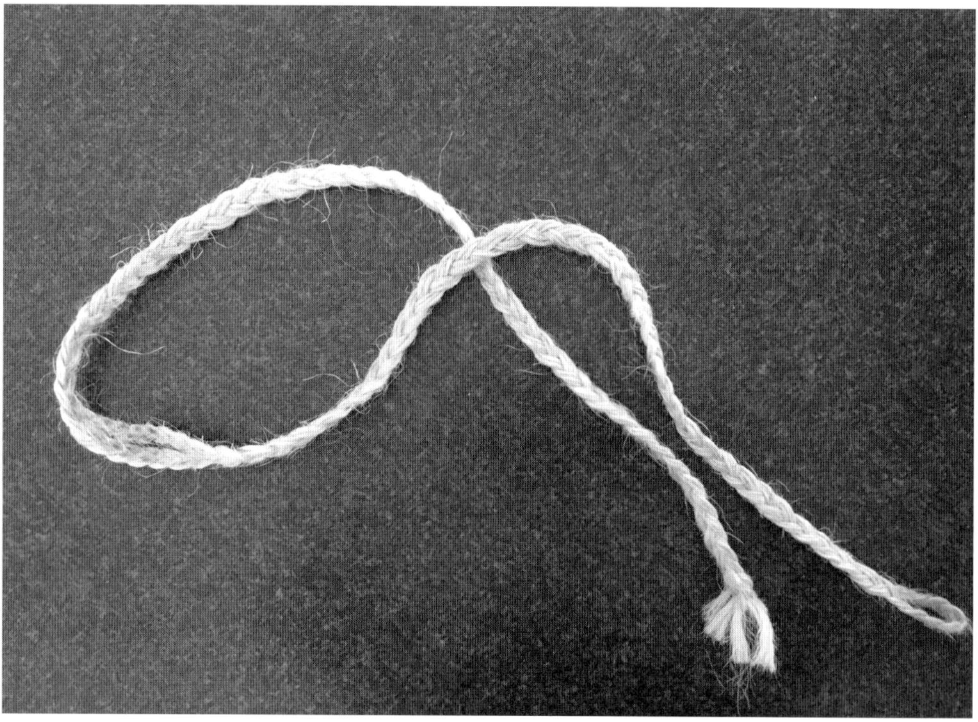

16. Balearic or split sling of flax. (*Author's collection*)

17. Clay slingshot from Carthage. (*Author's collection*)

18. Lead slingshot with image of a lightning bolt (left) and inscription 'Take That' (right). (*Author's collection*)

19. Reconstructed bronze Corinthian helmet before damage sustained. (*Author's collection*)

20. Reconstructed bronze Corinthian helmet after damage from 45g lead slingshot. (*Author's collection*)

21. The caved zygomatic region and frontal bone which was produced by the impact of a 120g limestone projectile. (*Author's collection*)

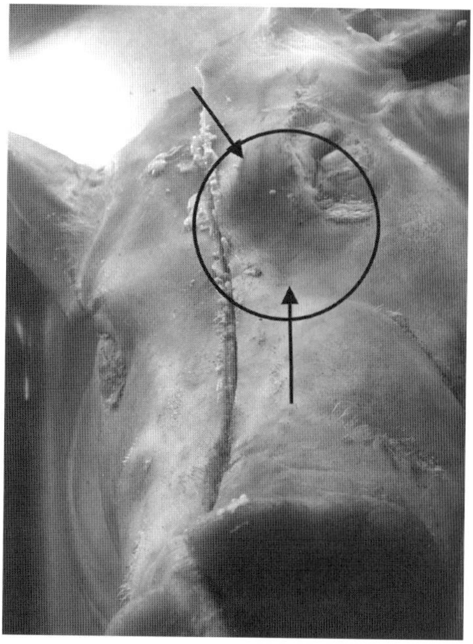

22. Reconstructed *akōn* javelin with *ankylē* throwing strap. (*Author's collection*)

23. Close-up view of throwing strap. (*Author's collection*)

24. Thracian peltast with dual-purpose spear for thrusting and throwing, a *pēlte* shield, high boots and battle cloak. (*Public domain*)

25. Reconstruction of a bronze *pēlte*. On the left, the wooden core can be seen, and on the right the bronze outer layer. (*Author's collection*)

26. Reconstruction of a forward-curved *machaira/kopis* sword (top) and a straight-bladed *xiphos* (bottom). (*Author's collection*)

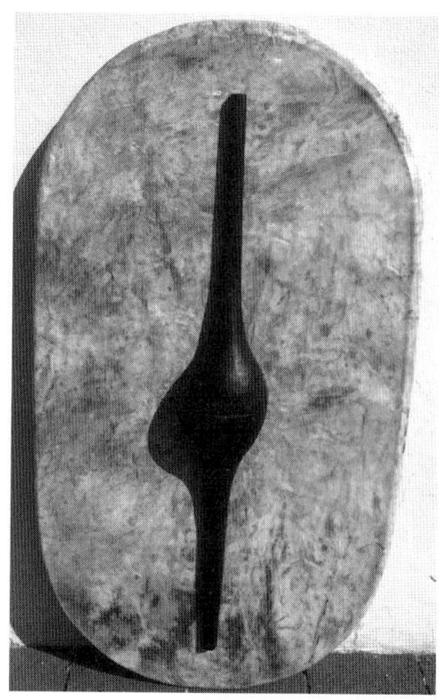

27. Recostruction of a *thureos* shield. (*Author's collection*)

28. Stele of Salamas, son of Moles of Adada, depicting him as a *thōrakitai*. (*Public domain*)

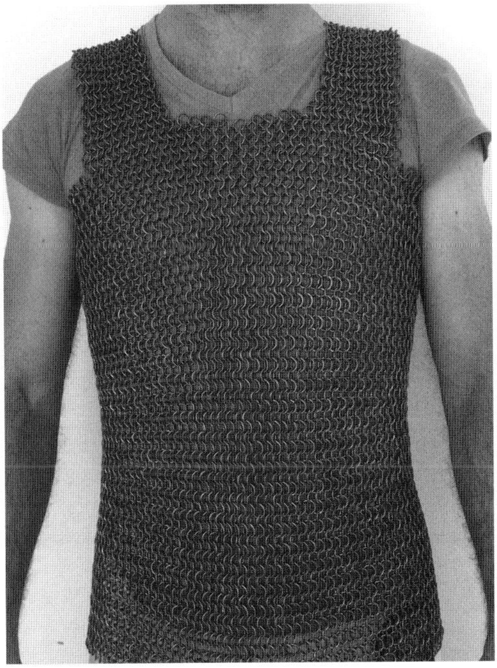

29. Mail reconstruction. The mail rings are butted in the same way as the third-century BC Ciumeşti find. The design reflects that of the soldier depicted in the Stele of Salamas, son of Moles of Adada. It is, however, possible that the Salamas mail was riveted as opposed to butted. (*Author's collection*)

30. Archer's crouching shot, with shield attached to forearm. (*Author's collection*)

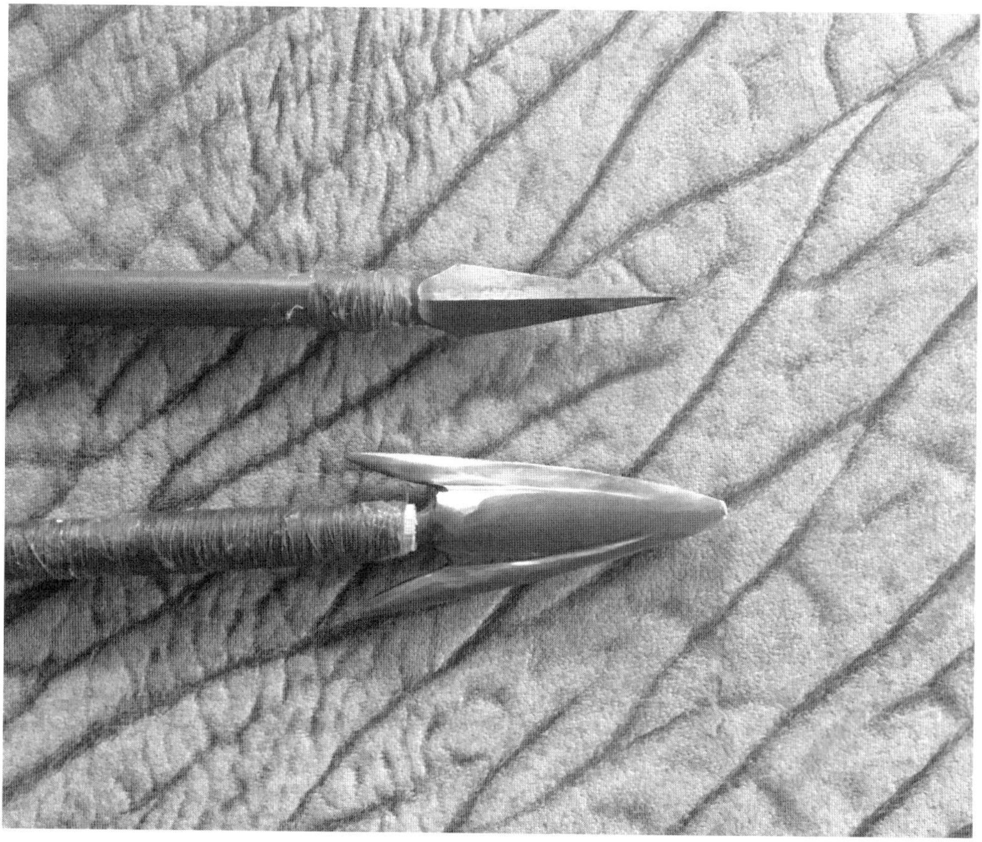

31. Reconstructed bronze Type IA 5 'Cretan' arrowhead, and 7g bilobate 'Scythian' arrowhead. (*Author's collection*)

32. Size difference between the 45g lead slingshot, the 120g stone and the *mina* (433g) stone. (*Author's collection*)

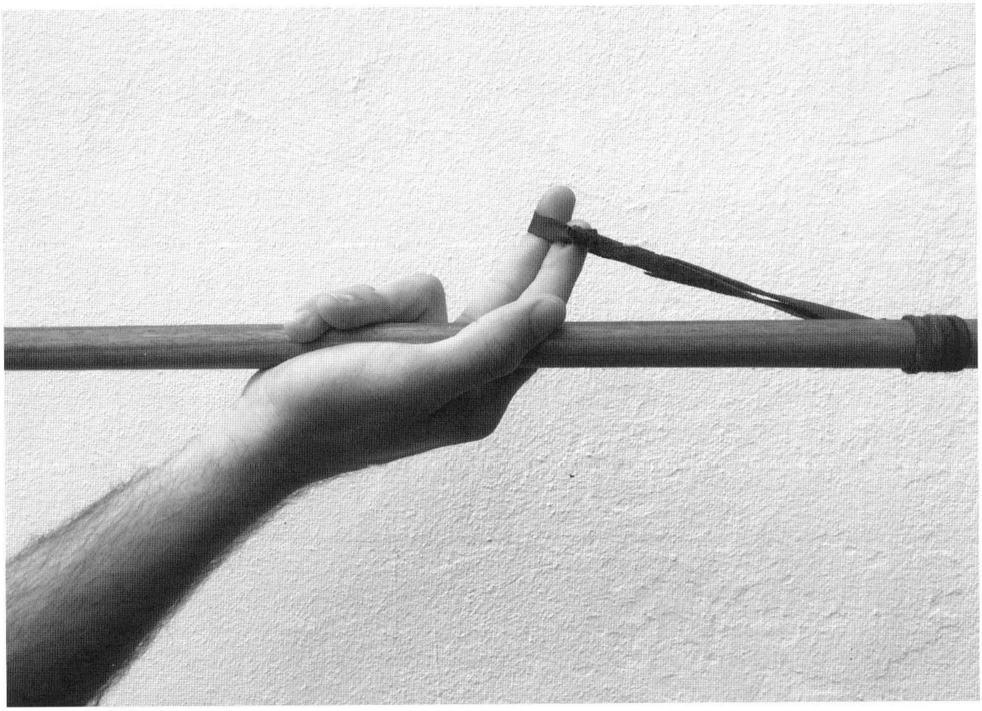

33. Method of holding the javelin with *ankylē* throwing strap. (*Author's collection*)

34. Length of javelin and close-up view of javelin head. (*Author's collection*)

35. Putty mould used to cast sling projectiles. (*Author's collection*)

36. The 45g lead projectile used for experimental archaeology. (*Author's collection*)

37. Slinging in action. (*Author's collection*)

38. Shooting long-distance parabolic shots. (*Author's collection*)

4.4 The Javelinmen and Peltasts in Combat

Akontistai

The early *akontistai* in their most basic form – those who were recruited from the poorest of society, servants of hoplites, oarsmen of a trireme and even slaves such as the Spartan *helots* – had the most rudimentary of equipment and training. Armed with only a few javelins, they were pure skirmishers, wearing no armour. Some individuals used cloaks or animal skins for protection, while others fought completely naked. Some of these *akontistai* were in all likelihood fairly skilled with the javelin, having learnt to wield it while hunting as the javelin was the primary hunting weapon in Greece and Rome. But other than the hunting of game, they were an organized mob at best. They would avoid melee engagement at all costs due to their lack of defensive equipment and training. Their job was to disrupt the enemy with their javelins before retreating when the opposition came close. They could be highly effective on the deck of a ship, as in the case of the Samothracian *akontistai* who completely cleared the deck of an enemy ship before commandeering it. They were also effective against heavy infantry without light-troop support of their own, as in the case of the failed Athenian expedition against Syracuse, where the Athenian hoplites were slowly worn down and utterly beaten by the Syracusan *akontistai*.

Peltasts

As we have seen, the growing need for javelinmen and the realization of their effectiveness during the Peloponnesian War led to the development of more professionalized light troops, which were better trained and equipped than the basic *akontistai*. Since they carried light *peltē* shields, they were able to defend themselves better than the basic *akontistai* from the missiles of other light troops, and also had some protection in a melee engagement. Being primarily armed with javelins or throwing spears meant they were first and foremost missile troops who fought from a distance. But since they were armed with secondary weapons such as swords, they could also close the gap and engage in a melee.

It was, however, their mobility that was their true strength. This, combined with better training and discipline, meant *peltasts* were perfectly adapted for special operations. With only the most elementary training and the bare minimum equipment, as well as their having low status in society, the basic *akontistai* could simply not be relied on for such specialized roles. *Peltasts*, on the other hand, having some training, an understanding of military matters, and the discipline and morale that this instilled in them, could be used for delicate missions such as Brasidas' capture of Torone. Because their light equipment caused them no hindrance and they could easily operate in rough and difficult terrain, they were ideally suited to performing forced marches and for the capture of strategic positions such as hills or mountain passes. While an isolated case in literature, the better training and discipline is reflected in Thracian *peltasts* from the Dii tribe, who had the ability to ward off cavalry. These Thracians were clearly trained and highly disciplined to be able to form a close formation and stand their ground against cavalry charges.

The mobility of *peltasts* also made them ideal for protecting the vulnerable heavy infantry while marching in column, especially against attacks from other light troops. Their equipment made them light enough to engage other light troops, either with missiles of their own or by closing the distance and engaging them in hand-to-hand combat. The Altıkulaç Sarcophagus depiction clearly shows the Greek mercenary *peltast* with his drawn sword, choosing to engage in a melee, despite having two remaining javelins. Being armed with shields and close-quarters weapons would give them an advantage over other *psiloi*, who fought only with missile weapons, when they engaged in a melee.

These agile *peltasts* were also able to take on the very best heavy infantry under the right circumstances. They could pelt the phalanx at a distance with javelins until they had significantly weakened and disrupted their formation. If the weighted-down heavy infantry attempted to charge them, the *peltasts* would simply retreat until no longer pursued. There was little hope of a heavy infantry soldier catching a retreating *peltast*. Furthermore, it was when the heavy infantry broke their formation to charge the *peltasts* or flee the field of battle that the *peltasts* were at their deadliest. Breaking ranks simply further exposed the heavy infantry to the missiles of the *peltasts*. As individuals, they would then be far more vulnerable than in their compact formation and could more easily be picked off by the *peltasts*.

Peltasts threw their javelins from fairly close range, somewhere between 10 and 20 metres. The main strengths of the javelin missile were accuracy and penetration power. Throwing a javelin at a range of more than 20 metres would require it to be thrown in a parabolic arc. As with archery, this would reduce accuracy as well as penetration power. As we have seen, the number of javelins the *peltast* could carry was limited. In order to use their javelins to maximum effect, they thus had to throw their javelins at a close range. Experimental archaeology has demonstrated that the javelin's effective range is 10–20 metres (see Javelin & Throwing Strap in Chapter 8). This is supported by modern game hunting, where the optimal hunting range with the javelin is between 10 and 15 metres.[79] If the target was armoured, one would want the javelin to hit the target with maximum kinetic energy. Once again, experimental archaeology has demonstrated that throwing a javelin with the *ankylē* throwing strap at 10–15 metres carried enough kinetic energy to penetrate period hoplite armour.[80]

Throwing their javelins from as close as 10 metres would allow the *peltasts* to pick individual targets from among the heavy infantry formation, cast the javelin with maximum penetrative power and still have more than enough distance between them and the heavy infantry to retreat if the heavy infantry sought to charge them. The heavily burdened hoplites would simply be unable to catch a lightly armed *peltast*.

The Battle of Lechaeum (391 BC) serves as a perfect example of the effective use of *peltasts* against heavy infantry. Xenophon paints a vivid picture in his *Hellenica* of these 'nimblest of all troops' who were able to defeat what had been considered to be the greatest of all Greek warriors:

> 'Iphicrates, leader of the peltasts, when they descried the Lacedaemonians and saw that they were not only few in number, but also unaccompanied by either peltasts or cavalry, thought that it was safe to attack them with their force of peltasts. For if they should proceed along the road, they could be attacked with javelins on their unprotected side and destroyed; and if they should undertake to pursue, they with their peltasts, the nimblest of all troops, could easily escape the hoplites … Iphicrates with his peltasts attacked the Lacedaemonian regiment. Now when the Lacedaemonians were being attacked with javelins, and several men had been wounded and several others slain, they directed the shield-bearers to take up these wounded men and carry them back to Lechaeum; and these were the only men in the regiment who were really saved. Then the polemarch ordered the first ten year-classes to drive off their assailants. But when they pursued, they caught no one, since they were hoplites pursuing peltasts at the distance of a javelin's cast; for Iphicrates had given orders to the peltasts to retire before the hoplites got near them; and further, when the Lacedaemonians were retiring from the pursuit, being scattered because each man had pursued as swiftly as he could, the troops of Iphicrates turned about, and not only did those in front again hurl javelins upon the Lacedaemonians, but also others on the flank, running along to reach their unprotected side. Indeed, at the very first pursuit the peltasts shot down nine or ten of them. And as soon as this happened, they began to press the attack much more boldly. Then, as the Lacedaemonians continued to suffer losses, the polemarch again ordered the first fifteen year-classes to pursue. But when these fell back, even more of them were shot down than at the first retirement. And now that the best men had already been killed … the Lacedaemonians themselves kept continually becoming fewer and fainter of heart, while their enemies were becoming bolder, and those who attacked them continually more numerous.'[81]

The battle was a devastating defeat for the Spartans, reinforcing the idea that they could be beaten in the field. Prior to the battle, the Boeotians were discussing peace terms with Agesilaus of Sparta, but the defeat encouraged them to abandon talks and continue with the war. It also brought an end to any significant Spartan operation in the territory of Corinth. It should be noted that the *peltasts* at the Battle of Lechaeum, even though they fought under Iphicrates, were not Iphicratic *peltasts*. His reforms allegedly took place after 374 BC, several years after the clash at Lechaeum, which took place in 391 BC.[82] The *peltasts* at Lechaeum were clearly javelinmen who refrained from close combat, whereas the Iphicratic *peltasts* were, as we have seen, primarily melee units.

Iphicratic Peltast

As discussed, the Iphicratic *peltast* was an attempt to find a compromise between heavy and light troops by equipping soldiers with lighter defensive gear and lengthening their offensive melee equipment. The reform was just as much a means to counter the financial burden required to equip heavy troops. As we have seen, those who became mercenaries did so because of desperation. The lighter equipment was far more financially viable than traditional hoplite gear. By equipping his mercenaries with the lighter gear, Iphicrates could deploy light infantry that could operate as heavy infantry without having the financial burden associated with fielding heavy troops.

Being equipped with linen cuirasses (*linothōrax*) and high boots (*Iphicratids*) gave them sufficient body defence. The cuirass was light and could easily flex to the wearer's movements. Since the metal cuirass and greaves had to be made and fitted to each individual wearer, and because of the material they were made of, they were very expensive. Linen armour and leather boots were far cheaper. Furthermore, the hoplite *aspis* shield was as expensive as it was heavy. Replacing it with the smaller, lighter and more economical *peltē* allowed the *peltast* to retain his mobility, yet have adequate and affordable protection in a melee and against missiles.

The Iphicratic *peltast*'s shield and body armour did not provide quite as much defence as the large *aspis* shield and metal *thōrax*, but their offensive equipment also played a role in the *peltast*'s defence. The Iphicratic spear was half as long again or even double the length of the hoplite *doru*, depending on the source. The long reach of the new spear provided the Iphicratic *peltast* with additional defence. Light troops had always relied on keeping a safe distance from their foes through the use of missile weapons, and while the Iphicratic *peltast* still relied on distance for defence, this was now in the context of the melee with a weapon that had tremendous reach.

The Iphicratic *peltast*'s equipment was still light enough for him to operate in difficult terrain and perform specialist operations such as the capture of strategic positions. They may in some cases have struggled using their long spears, but both Nepos and Diodorus mentioned the lengthening of their swords. In the event of the long spear becoming too much of a hindrance, the *peltast* could now switch to his sword. As both sources on the Iphicratic *peltast* emphasized the adoption of longer swords, it is clear that they had become an important part of their fighting style, no longer merely a secondary side-arm.

Whereas light troops had generally operated primarily as missile range troops, the Iphicratic *peltast* reforms created light troops mainly focused on the melee. They were more cost-effective to deploy, and because of their tremendous reach advantage were able to meet and engage heavy infantry in the field. I am in agreement with the scholars who make the argument that the Iphicratic *peltast* was the predecessor of the Hellenistic pike phalanx of Philip II and Alexander the Great. If the accounts of Nepos and Diodorus have any merit, it is hard to deny the uncanny similarities between the Iphicratic *peltast* and the Hellenistic phalanx. The fact that Nepos and Diodorus claim Iphicrates won great fame for these innovations and improvements more than suggests that this was a very successful reform. Outside of these two accounts, however, there is no mention of the Iphicratic *peltast*. If it was as successful as the sources claim, why is it not mentioned in other sources? And why did the unit disappear from the battlefield? Perhaps the truth is that the reforms were successfully adapted as the pike phalanx of the Hellenistic era.

4.5 Conclusion on Javelinmen and Peltasts

The magnitude, both in scale and duration, of the Peloponnesian War and the conflicts thereafter emphasized the limitations of Greek heavy infantry and the need for light troops. The result was better-trained and equipped light infantry, predominantly armed, for practical reasons, as javelineers. It was more sensible to arm mercenaries and those who could not afford heavy infantry gear with javelins as opposed to bows and slings, which required a higher level of skill and specialization. Javelins were also much more cost effective than archery equipment. It was also the case that most males in Graeco-Roman society would have had some experience with the javelin through hunting and athletics, making the skill more easily transferable to warfare. Because of this, it was more cost effective to deploy javelinists, meaning more fighting men could be equipped and deployed, swelling manpower for campaigns.

The ever-growing reliance on light troops saw the natural evolution of the light infantry to fill roles beyond skirmishing and range warfare, culminating in the Iphicratic *peltasts*, who retained the mobility and nimbleness of light troops but were predominantly armed as melee infantry and had the ability to engage heavy infantry at close range. As discussed, the effectiveness of the Iphicratic *peltast* was in all likelihood the precursor of the Macedonian-style phalanx which dominated Hellenistic warfare for more than two centuries.

Chapter 5

The Multi-Purpose Hybrids of the Hellenistic Era

The larger-than-life figure of Alexander the Great and his conquest of vast territories ushered in a new era known as the Hellenistic Age. Upon his death, his empire was carved up among his generals, who ruled as monarchs of their respective regions. The idea of uniting Alexander's empire was quickly lost as these monarchs set up dynasties of their own. It did not take long for conflict to erupt between these Successor states, with unprecedented large-scale wars and blood feuds which lasted for generations. The period was marked by revolts, civil wars and usurpations as the rulers fought among themselves and desperately sought to hold onto their kingdoms, which they ruled as foreign monarchs over an often-hostile native population. Furthermore, the period saw the rise of Rome from a fledgeling city-state to a colossal empire through countless wars of its own, which eventually brought it into conflict with the Hellenistic kingdoms themselves. Rome in due course swallowed up these Hellenistic kingdoms, along with the rest of the Mediterranean world, creating one of the world's most enduring empires.

Warfare was thus central to the Hellenistic era and the societies who lived therein, forming the foundation for survival and prosperity. Conflict was extremely frequent and continuously shaped the Hellenistic world. It is thus not surprising that warfare was constantly moulded and reshaped throughout the era. The age saw a definite move to a more diverse and multi-purpose light infantry that could both fight from long range and actively engage in melees, thereby playing a more active role in pitched battles.

Hellenistic Elite Guard Peltast

The gap between light and heavy troops was to some extent filled by Alexander's Successors in the form of the elite Hellenistic *peltasts*. The term *peltastai* appears numerous times in the *Histories* of Polybius. However, whenever Polybius uses the term *peltasts* within the armies of the Successor Kingdoms, it no longer refers to a second-class mercenary soldier but to the elite royal guard infantry of Ptolemaic Egypt, Antigonid Macedonia and the Seleucid Empire.

The guard infantry *peltasts* – the elite infantry of the great Hellenistic kingdoms – served as a standing force. They were conscripted from across their respective empires and recruited for their youth, vigour, strength and skill.

These elite *peltasts* could operate as heavy infantry, fighting in disciplined phalanx formations in large-scale pitched battles, but were also able to deploy in difficult terrain, storm breaches in sieges, conduct forced marches, act as assault infantry and stage ambushes.

These elite guard infantry *peltasts* of all three Successor Kingdoms are at some point described as fighting in the Macedonian phalanx formation. Both the Ptolemaic and Seleucid *peltasts* deployed as heavy infantry in phalanx formation at the Battle of Raphia (217 BC). At Raphia, the Seleucid *peltasts* – better known as the *argyraspides*, or Silver Shields – are explicitly described by Polybius as armed in the 'Macedonian manner'.[1] The Ptolemaic *peltasts*, while not directly referred to as fighting in phalanx formation, are mentioned by Polybius as training alongside the Ptolemaic phalanx and at the battle itself fought in the main battle line alongside the Ptolemaic phalanx.[2] Similarly, the Antigonid *peltasts* under Philip V fought in phalanx formation at Cynoscephalae (197 BC) on the right flank of the Macedonian citizen phalanx.[3] The Antigonid *peltasts* are also described by Livy as fighting in the phalanx and armed with *sarissa* pikes at the Battle of Pydna (168 BC).[4] Polybius describes the Antigonid *peltasts*, when attacked by Aetolian cavalry, as able to form up and fight in the *synaspismos* formation, the most compact phalanx deployment where the soldiers interlocked their shields to form an impenetrable wall of bronze.[5]

As mentioned, these *peltasts* were capable of operating in difficult terrain. Indeed, when Philip V moved to capture Thalamae, he left behind his baggage train and most of his army and took with him only his light troops and *peltasts*. This was because of the difficult terrain, Polybius writing that 'the approaches to it [Thalamae] are narrow and difficult and the place itself secluded and not easily entered'.[6] *Peltasts* were also used for forced marches. In an attempt to catch the retreating Aetolian army of Phillidas, Philip V left most of his army behind and took only his light troops and *peltasts* on a forced march to overtake the enemy (218 BC).[7] Similarly, Antiochus the Great left the majority of his forces behind and force-marched his cavalry, light troops and *peltasts* to capture the ford at the Arius River (208 BC).[8] The *peltasts* of both the Seleucids and the Antigonids were used as assault troops in sieges to attack breached walls. The *peltasts* of Philip V stormed the breach of Pale in 218 BC, and Antiochus' *peltasts* poured through the breach at Sirynx in 209 BC.[9] The *peltasts* were capable of ambushes, as those of Philip V clearly demonstrated when they were deployed along with some Illyrians in a successful ambuscade against the Aetolians.[10] *Peltasts* were also used to guard and protect the heavy infantry of the phalanx, who were vulnerable on the march. During the Social War (220–217 BC), Philip V used his *peltasts* and cavalry to defend his marching phalanx from a Spartan attack.[11] According to Polybius, the *peltasts* gave an excellent account of themselves, 'displaying great gallantry'.[12]

What we can take from this is that the need for light infantry had become so essential that the great Hellenistic empires took their very best infantry soldiers and trained them to fight as a type of hybrid infantry, capable of fighting in the phalanx of the heavy infantry but also as a type of light infantry. As mentioned at the beginning of the book, large-scale pitched battles were relatively rare at this time. Campaigns were dominated by countless skirmishes, long marches where columns of heavy infantry and baggage trains needed the protections of light infantry, operations in difficult terrain, the capture of strategic locations, ambushes, forced marches and drawn-out sieges. This was not the domain of the heavy infantry. Warfare had become increasingly focused on the elements where the heavy infantry could not operate, meaning the role of light infantry became ever more important. The growing importance of light infantry is perfectly reflected in the infantry guard *peltasts* of the Hellenistic empires.

5.1 Arms and Armour of the Guard Peltast

5.1.1 Heavy Armed Peltast

Sarissa

From the literature, it is clear that the elite guard *peltasts* of the Hellenistic Successor states were able to be fielded in the main battle line as heavy infantry. They were armed with the monstrous *sarissa* pike that ranged in length from 10–16 cubits, depending on the period. The cubit during the Hellenistic era was based on the Olympic/Peloponnesian cubit of 48cm, as opposed to the Attic cubit of 45cm.[13] From this, the Hellenistic *sarissa* ranged from 480–768cm.

During the time of Philip and Alexander, the *sarissa* was between 10 and 12 cubits (480–576 cm).[14] Polybius, who served as a military officer of the Achaean League, was undoubtedly an eyewitness of the 'Macedonian' phalanx, which had dominated Greece and the Eastern Mediterranean for the better part of two centuries. In his description of the *sarissa*, he states that the length of the pike 'is sixteen cubits according to the original design, which has been reduced in practice to fourteen'.[15] What we can take from this is that the *sarissa* was lengthened from 12 cubits (576cm) to eventually 16 cubits (768cm) in an attempt to retain a reach advantage over one's enemies. The 16-cubit *sarissa* is also mentioned by Polyaenus and Aelian.[16] It seems, however, that the *sarissa* of 16 cubits became impractical, and by the time of Polybius it was reduced to a more reasonable 14 cubits (672cm).[17]

The *sarissa* was heavy and unwieldly, so much so that Polybius comments on the Greeks suffering under its weight and cumbersome length: 'For the Greeks find it difficult to hold even their *sarissae* on the march and can scarcely bear the fatigue of them.'[18] It was thus not suited for the various tactics and combat style usually associated with light infantry. As we have seen, the guard *peltasts* were constantly operating as light infantry in roles outside of pitched battles, such as forced marches, ambushes, fighting in broken terrain, guarding the heavy infantry on the march and assaulting breached walls during sieges. It is highly unlikely that they were armed with the unwieldy *sarissa* when fielded in such roles, and we may assume that the *sarissa* was reserved for the static line battles dominated by the heavy infantry.

Macedonian Peltē

Because the *sarissa* was a weapon of such immense size, it had to be wielded with both hands. A smaller shield than the old *aspis* was thus needed; as their name implies, the guard *peltasts* now carried the *peltē* shield (**see plate 24**). The Macedonian *peltē* was attached to the left forearm, which allowed for the *sarissa* to be held with both hands. Asclepiodotus describes the shield as follows: 'The best shield for use in the phalanx is the Macedonian, of bronze, eight palms in diameter, and not too concave.'[19] This is mirrored by Aelian's writing: 'The Macedonian shield, made of bronze, is best. It must not be too concave and should be eight palms in diameter.'

Based on the Olympian/Peloponnesian cubit, the Greek palm was 8cm, giving the Macedonian *peltē* a diameter of 64cm. This is backed by several archaeological finds, such as a fragmented *peltē* shield from Begora with an estimated diameter of 66cm.[20] A shield find at Pergamum is slightly more intact and measures between 65 and 66cm.[21] The fascinating find of a Ptolemaic shield mould of limestone, now in the Allard Pierson Museum in Amsterdam, has a diameter of approximately 69cm.[22] At first, it appears as if the mould could not be that of a *peltē* as the diameter is too large. The Pergamum shield find, however, clearly shows that the outer layer of the shield was folded over around the wooden core of the shield.[23] This folded section is calculated to be approximately 2.5cm.[24] The folding of the outer layer is confirmed by the shield of Pharakes of Potus, which shows the outer rim cut into wedges meant to fold around the wooden core.[25] When the folded outer layer along with the thickness of the shield is taken into consideration, the Ptolemaic mould corresponds nicely with the other shield finds. The archaeological evidence thus supports Asclepiodotus and Aelian when they say the *peltē* is eight palms in diameter.

Body Armour

There is very little information on the armour of the guard *peltasts*, but we may assume that they wore some form of body armour when fighting as a phalanx in pitched battle. The ordinary citizen phalanx certainly wore body armour, according to the Amphipolis Decree.[26] The front-rank soldiers were clad in metal *thōrax* armour, while the ranks that followed wore the *kottybos*, which is thought to be the *linothōrax* (see Chapter 2.1.3, Defensive Equipment: Body Armour). In addition to the body armour, they wore greaves (*knemides*) and a conical *konos* helm. These troops were also armed with a short *makhaira* sword as a secondary weapon.

If the ordinary citizen phalanx was clad in body armour, it is reasonable to assume that the elite guard unit, when fighting in pitched battles in phalanx formation, also wore armour. As the elite standing infantry body guard unit, they were supplied by the state and their equipment would in all likelihood be of the highest quality. One can imagine their gear to be gilded and embossed with decoration. We do know that the Seleucid guard *peltast* carried silver-plated shields, hence their name *argyraspides* (Ἀργυράσπιδες), which translates as Silver Shields. The *agema*, an elite unit among the Antigonid and Ptolemaic *peltasts*, is described at the Battle of Pydna (169 BC) as 'gleaming with gilded weapons'.[27]

5.1.2 Light-Armed Guard Peltast

Weapons

As discussed, the *sarissa* was too long and cumbersome to use in typical light troop tactics. It would simply not have been possible for these elite *peltasts* to use the monstrous pike in most of the light troop roles. We thus have to assume that when they were acting in light troop roles, the guard *peltasts* were armed with other types of weapons. The spear had always been a popular weapon among the Greeks, and it is very possible that they opted for a short spear when not acting as heavy infantry. There is evidence for the *peltast* being armed with a spear when not fighting in a phalanx formation. The mural from the 'House of Menander', depicting the fall of Troy, is a copy of a third-century BC original which depicts Ajax the Lesser dragging Casandra away from a statue of Athena.[28] Carrying a *peltē* shield, Ajax is believed to represent an elite guard *peltast*.[29] As a primary weapon, Ajax is armed with a thrusting spear approximately 2 metres in length. Hanging from Ajax's left hip is a dagger, or possibly a sword. Swords were becoming increasingly popular among the Greeks during the Hellenistic age, so much so that the mock combat between two fighters using swords and shields known as *thureomachia* became a popular athletics event during the period.[30] These types of weapons were far more suitable for light infantry action than the cumbersome *sarissa*.

Yet another image which has been interpreted as an elite guard *peltast* is the funeral stele of Zoilos dating from the early second century BC.[31] Zoilos bears a *peltē* shield on his left arm, while in his right hand he holds what appears to be a short spear. Some scholars have interpreted this as a *sarissa* because of its length, protruding beyond the boundaries of the image. However, upon closer inspection one can see that the shaft culminates in a spearhead, which supports the idea that guard *peltasts* were armed with short thrusting spears.

Peltē

As we have seen, the relatively small size and lightness of the *peltē* made it ideal for light infantry action (**see plate 25**). While the guard *peltasts* would discard the heavy *sarissa* for operating outside of the phalanx, there was no need to exchange their *peltai* for other types of shields as the *peltē* was suited for both heavy and light infantry roles. Both Ajax and Zoilos bear what has to be interpreted as *peltai* shields.

Body Armour

Ajax is depicted without any form of body armour, with the exception of a *pilos* (πῖλος) cap. The *pilos* was a thick conical cap made of felt (the word '*pilos*' referring to the method of compressing wool to make felt) that was usually worn under a metal helm for padding. Such a padded felt cap was light, not cumbersome, and did not restrict vision or hearing, yet still provided some protection for the head. It is impossible to tell if this was common practice, and there are scholars who interpret a passage from Xenophon that the Spartans at the Battle of Sphacteria (425 BC) were not wearing metal helmets but only had the *pilos* felt cap for head protection:

> 'The Lacedaemonians had now to sustain a rude conflict; their *piloi* would not keep out the arrows, darts had broken off in the armour of the wounded, while they themselves were helpless for offence.'[32]

It is possible that the *pilos* to which Thucydides referred was the metal helmet also called a *pilos*, having taken its shape from the felt cap. The helm became popular due to its ease of manufacture, and because it was an open-faced helmet it allowed for better vision and hearing. The fact that the helm was unable to 'keep out the arrows' is possibly due to the fact that the face was far more exposed in comparison to the older Corinthian closed-faced helm or the Attic and Chalcidian helms with their large cheek pieces.[33]

As with Ajax, Zoilos wears a *pilos* or *konos*. Due to the lack of colourization on the stele, it might, as in the case of Ajax, be a simple felt *pilos* or the metal version discussed above.

While Ajax in the depiction wears no other form of body armour, it is still possible that the elite guard *peltasts* operating outside of the phalanx wore other forms of light body armour, such as a *linothōrax/kottybos*, which would not restrict their mobility. In fact, the stele of Zoilos seems to depict the soldier in what appears to be a cuirass with *pteruges*, indicating that he is wearing some form of body armour.

From the image of Ajax and Zoilos, and the description of their tactics in the literature, it is clear that the guard *peltasts*, when not fighting in the phalanx, were lightly armed and armoured in order to fill their role as light infantry.

5.2 The Elite Guard Peltasts in Combat

The Seleucid guard *peltasts* are described by Polybius as 'a force of ten thousand selected from every part of the kingdom'.[34] At the Battle of Raphia (217 BC), they are referred to as the 'picked Syrians', and at Magnesia as the 'royal cohort'.[35] The Antigonid *agema peltasts* are described by Plutarch (*Aem.*, 18.3) as 'picked men, the purest of the Macedonians on account of their virtue and age', and by Livy as chosen 'for their strength and the enduring energy of their age'.[36] These were the very best men that the Successor states could field as a permanent infantry force. The numerous skirmishes and tactical manoeuvres, such as forced marches and capture of strategic location, which led up to pitched battles, had become immensely important and often greatly influenced the outcome of these major clashes. It thus made a lot of sense for the Successor states to have their very best soldiers, those who were the best trained and equipped and the most disciplined, to fill important light infantry roles.

As heavy infantry, they fought in the Macedonian-style phalanx during large-scale battles. During these set-piece battles, they were often deployed on the right flank of the main battleline, which was considered the position of honour in keeping with their status of elite troops. They were highly trained and disciplined to fight as a cohesive unit, which took hours of drill and practice to perform the various formations.

As an elite standing infantry force, their discipline made them ideal for a variety of specialist roles.

King's Standing Infantry Guard

The backbone of the Hellenistic kingdoms was the citizen soldier of the Macedonian-style phalanx. These citizen soldiers were not professional soldiers in the modern sense of the term, but were land-holding farmers. In exchange for their farming land, they served in the citizen phalanx. These farmer-soldiers could not be away from tending to their land for very long periods, and were only mustered in times of need and during the campaigning season.

The guard *peltast* unit, however, was the closest thing the ancient world (prior to Rome's Marian Reforms of the late second century BC) had to a professional army. These men were conscripted to serve as a standing infantry force and were not disbanded at the end of the campaign season. During their time of conscription, they had no other obligations besides being soldiers. These men were thus always at the king's disposal, ready to march out to do their monarch's bidding at a moment's notice. Being a standing force meant the king had a permanent body of soldiers he could direct for a variety of roles, such as general policing, peacekeeping and guard duties. On campaign, the king had a well-trained and highly disciplined unit to which he could entrust delicate and important duties. As a standing force, they could campaign for much longer and far beyond the borders of their kingdoms, as their only responsibility was as a military unit. The guard *peltasts*, or the 'royal cohort' as Livy describes them, demonstrated their close proximity to the king as his elite guard infantry unit. They accompanied the king in person on every campaign and were present at every major battle, where they deployed close to their monarch.

Most light troops during the Hellenistic Period were still recruited from the poorer lower classes, levies and mercenaries. They were thus not always the most reliable of troops, being poorly trained and with little loyalty to the state. The guard *peltasts*, however, served the Hellenistic kings as highly trained and dependable light troops who were loyal to their monarchs. They could thus be relied upon during the most dire of situations in battle and in the most sensitive and delicate of tactical missions.

Forced Marches

Military campaigns often required troops to force-march to reach a specific location in a short period of time or to outmanoeuvre an enemy army. Many of these forced marches were conducted through broken and difficult terrain. The cumbersome equipment of the heavy infantry simply did not allow them to force-march long distances. The literature is rife with references of Hellenistic monarchs leaving behind the heavy infantry when setting forth on forced marches. The king or general thus had to rely on his light troops and cavalry to make these forced marches. Again, the light infantry, being mercenaries or conscripted levies, could not always be depended upon. Having a reliable, well-trained and equipped light infantry in the form of guard *peltasts* who could accompany the king on these forced marches was thus a valuable asset for any military commander.

A prime example is Antiochus the Great's forced march to rapidly capture a river crossing. Polybius tells us:

> 'The river being at a distance of three days' march, he marched at a moderate pace for two days, but on the third day he ordered the rest of his army to break up their camp at daylight while he himself with his cavalry, his light-armed infantry, and ten thousand peltasts advanced during the night marching quickly.'[37]

Melee Support and Assault Troops

The guard *peltasts*, even when acting as light troops, were predominantly melee units. Most light troops were skirmishers who focused on range combat, such archers, slingers and javelinmen, as discussed above. Being able to deploy highly trained and equipped *peltasts* in support of range light infantry allowed for effective combined arms tactics. This was often lacking in skirmishes and when operating in difficult terrain, since the heavy-armoured melee troops could not deploy under these circumstances. While some standard light troops such as the *peltasts* of the classical period could at times act as light melee units, their primary role was still range combat. In contrast, the guard *peltasts* were dedicated melee units who could act as support for the light infantry.

We do see in the literature that the *peltasts* often acted as a supporting force for other light infantry. An example would be Philip V's mercenary light troops engaging the Spartan forces outside Menelaïum. The battle took place in the hills and Philip left his heavy infantry behind. Crossing a river with his Illyrians, *peltasts* and mercenaries, Philip first sent his mercenary light infantry to attack the Spartans, who occupied the high ground. At first, Philip's mercenaries were getting the worst of the fighting due to the Spartans' elevated position and their being better armoured. Nevertheless, the battle swung in favour of the Macedonians when Philip sent in his *peltasts* to support the mercenaries and his Illyrians to outflank the Spartan position:

> 'Philip, on approaching Lycurgus, first sent the mercenaries alone against him, so that at the beginning the Lacedaemonians fought with more success, favoured as they were in no small degree by the ground and their heavier armour. But as soon as Philip had sent the peltasts to take part in the fight acting as a supporting force, and getting round the enemy with his Illyrians charged them on the flank, his mercenaries encouraged by this

> support fought with much greater spirit, while Lycurgus' force dreading the charge of heavier troops gave way and ran.'[38]

When Philip V campaigned against the Illyrians, we again see the light troops being supported by the elite *peltasts*. During a battle amongst the hills surrounding the cities of Lissus and Acrolissus, Philip sent his light troops to engage the Illyrians who occupied the hilltops. After fierce fighting, they were driven back due to the sheer number of the Illyrian warriors and having to fight an uphill battle. Upon fleeing, they were hotly pursued by the Illyrians, but were able to find refuge with the elite *peltasts* who were formed up in the plain below the hills. The discipline and training of the *peltasts* can be seen as they fought a retreating battle, slowly giving ground to the enemy and giving the impression that they were being beaten back. When the enemy had been lured far enough from their fortified cities, Philip sprang an ambush of light troops whom he had hidden among the wooded ravines. At the same time, the *peltasts* halted their feigned retreat and fell upon the Illyrians:

> 'Philip's troops [light-armed infantry], yielding to the difficulties of the ground and to superior numbers, were put to flight. When they took refuge with the peltasts, the Illyrians from the town in their contempt for them followed them down the hill and engaged the peltasts on the level ground. At the same time the garrison of Acrolissus, seeing that Philip was slowly withdrawing his divisions one after the other, and thinking that he was abandoning the field, imperceptibly let themselves be enticed out owing to their confidence in the strength of the place, and then abandoning Acrolissus in small bodies poured down by bye-paths to the level ground, thinking there would be a thorough rout of the enemy and a chance of some booty. But at this juncture the troops which had been posted in ambush on the land side rose unobserved and delivered a brisk attack, the peltasts at the same time turning and falling upon the enemy.'[39]

The successful ambush killed many of the garrison troops and demoralized those who survived to such an extent that Acrolissus 'was taken beyond all expectation at once and without striking a blow'.[40] Lissus was taken the very next day 'after a desperate struggle, the Macedonians having delivered several energetic and terrific assaults'.[41] There is little doubt that these 'terrific assaults' upon the city walls were spearheaded by the elite *peltasts*. The role of the guard *peltasts* as assault troops during sieges is testified in the literature. At the siege of Pale (218 BC), the guard *peltasts* of Philip V stormed the breached wall:

'[H]e set fire to the props and brought down all that part of the wall which had been underpinned, upon which he first of all sent forward the peltasts under Leontius, drawing them up in cohorts and ordering them to force their way through the breach.'[42]

During Antiochus' *anabasis* of reconquering the eastern provinces, the Seleucid army besieged the Parthian army of Arsaces II, which was shut-up in the city of Sirynx in 209 BC.[43] Polybius gives a brief account of the gruelling siege of the well-fortified city:

'The operation consisted chiefly of mining under fortifications. For the city was defended by three trenches, thirty cubits broad and fifteen deep, with a double vallum on the edge of each; and behind these there was a strong wall. Frequent struggles took place at the works, in which neither side were strong enough to carry off their killed and wounded: for these hand-to-hand battles took place, not above ground only, but underground also in the mines. However, owing to the numbers employed and the activity of the king, it was not long before the trenches were choked up and the walls were undermined and fell.'[44]

When the walls finally fell, it was the Seleucid guard *peltasts* who spearheaded the assault on the breach and captured the city: 'When the peltasts now vigorously forced their way through the breach, they surrendered in despair.'

We also see the Antigonid guard *peltasts* heading the assault against the fortifications of the city of Oaeneum in 169 BC. Perseus, the Macedonian king, sought to capture the city, which was an ally of Rome. The city was taken by the use of a siege mound, which was spearheaded by the 'royal cohort' (*cohors regia*):

'As soon as the mound was brought close to the wall, the royal cohort (the men of which are called Nicators) rushed from it into the town, while an assault was made by scalade in many places at once. All the males, who had reached the age of puberty, were put to the sword, their wives and children were thrown into confinement, and everything else was given as booty to the soldiers.'[45]

While Livy does not refer to the *peltasts* directly, scholars agree that the 'royal cohort' refers to the guard *peltasts*.[46] We have already seen that Livy referred to the Seleucid *peltasts* as the royal cohort, which supports the idea that the 'royal cohort' at the siege of Oaeneum were *peltasts*.[47] Bearing the nickname 'Nicators', meaning 'conquerors', speaks of the skill and elite status of these troops within the Antigonid army.

5.3 Conclusion on the Hellenistic Peltasts

The importance of having a professional standing army which was well equipped, and highly disciplined and trained, cannot be emphasized enough. They could operate not only as heavy infantry in pitched battles, but also as light infantry. Their duality allowed for them to be fielded as the very best of the heavy infantry during pitched battles, yet be elite light infantry under a variety of other circumstances. Consequently, they became the right hand of the king, always at his side, taking the field against his enemies, whether in large pitched battles, as support troops, during sieges or when rapid forced marches were required.

Thureophoroi and Thōrakitai

From the third century BC, a new troop type, the *thureophoroi*, started to appear in Hellenistic armies. The *thureophoroi* took their name from the type of shield they carried, which was called a *thureos*. By the third century BC, the era of the heavily armed hoplite in mainland Greece had come to an end, and the struggling economies of many of the smaller Greek states had adopted *thureophoroi* as their standard infantry unit. These troops were a light dual-purpose troop type adapted for skirmishing as well as close combat fighting, which better suited the smaller states' needs of guerrilla warfare, border protection, operations in rough terrain and a more economical approach to arming their troops. This new troop type quickly replaced the classical *peltasts* as the main skirmisher force of the Greek-influenced world.

Greek mercenaries from the third to the first century BC are especially regarded to be *thureophoroi*.[48] The economic struggles which the smaller Greek states faced in the shadow of the Hellenistic kingdoms meant that Greece was an ever-present source of mercenaries, the continuous warfare throughout the period providing coin and opportunity for a man willing to pick up the spear. Common cultural affiliations, such as language, also played a role in the popularity of Greek mercenaries being hired by the Successor states. Many Greek mercenaries served as *symmachoi*, or contracted mercenaries. These mercenaries served in foreign armies based on treaties between two states, people or even cities, which allowed the treaty members special recruiting rights within the other's sphere of influence.[49] Greek mercenaries were hired from different areas across the mainland, and many of the hired soldiers would not have known each other personally and thus would not have had the opportunity to train together. The *thureophoroi* style of fighting did not require the unison and countless hours of drilling which the heavy infantry of the phalanx did. It also made sense that Greek mercenaries, in general, were armed as *thureophoroi*, since it had become the fighting style of their homeland and was thus a familiar style of warfare to them. Because mercenaries did not need to be disbanded outside of the campaign season, they were ideally suited for garrison duties.[50] The *thureophoroi* fighting style was also far better suited for garrison duties.

There is little direct reference to the *thureophoroi* in literary sources. The term appears only a handful of times in Polybius' *Histories*, which serves as arguably the most important military literary source of the third century BC.[51] *Thureophoroi* also appear in Plutarch, who states that the Achaean army of the third century was armed as *thureophoroi*, but under Philopoemen they were rearmed and trained to fight in the phalanx.[52]

As mentioned, Greek soldiers and mercenaries were generally armed as *thureophoroi*, and it is very reasonable to assume that the majority of Greek soldiers referred to in the literature were *thureophoroi*. The Aetolians who had become famous, and at times infamous, for their mercenary service during the Hellenistic period would have served as *thureophoroi*. The Greek mercenaries of the Seleucids at Raphia also served as *thureophoroi*.[53] So too were the mercenaries Philip V used in his campaign against Lycurgus of Sparta.[54] It is interesting to note that these mercenaries of Philip are later in the same passage described as *euzōnoi* (see Chapter 1.2.2, Euzōnoi). Similarly, the Spartan and Achaean mercenaries at the Battle of Mantinea are also mentioned as *euzōnoi*.[55] It is thus very reasonable to assume that most *euzōnoi* were *thureophoroi*.

While there might not be plenty of references to *thureophoroi* in literature, we do have an abundance of iconographical and archaeological sources of *thureophoroi*, particularly from funeral stelae of mercenaries in Ptolemaic or Seleucid employ.[56]

The fighting style of the *thureophoroi* was so prominent and widespread that it became the popular athletics event known as *thureomachia*, where participants engaged each other in mock combat, armed as *thureophoroi*.[57]

5.3 Arms and Armour of the Thureophoroi

Plutarch gives a fair description of the arms of the Achaean *thureophoroi* before being retrained as phalangites:

> 'In the first place, however, he [Philopoemen] changed the faulty practice of the Achaeans in drawing up and arming their soldiers. For they used shields [*thureoi*]; and spears which were much shorter than the Macedonian pike [*sarissa*]. For this reason they were effective in fighting at a long distance, because they were so lightly armed, but when they came to close quarters with the enemy they were at a disadvantage.'[58]

Plutarch's description of *thureophoroi* corresponds rather well with that of the iconographical sources, and we can thus accurately reconstruct the *thureophoroi* along with their arms and armour.

5.3.1 Thureos

The *thureos* shield, from which the *thureophoroi* derived their name, translates as 'door' in ancient Greek, reflecting its oblong/oval shape (**see plate 27**). The *thureos* was in all likelihood adopted following the Gallic invasions of the 270s BC, after fighting against the Gauls who used similar types of shields. Another possibility is that they were introduced by Pyrrhus of Epirus returning from his Italian campaigns.[59] Whatever the case may be, the use of the *thureos* quickly spread throughout the Hellenistic world and beyond, being widely attested in iconographical evidence and referred to in the period literature.[60]

The best-preserved and best-known iconographical evidence of *thureophoroi* is probably the Sidon Stelae collection.[61] The depictions show the shield to be oval in shape and, as per Plutarch's description, rather narrow, while many revealing the length to be from the shoulder to just below the knee.[62] The surface of the shield appears flat or with minimal curvature. Plutarch writes that the shields were 'easily carried because they were so light'.[63] From this we may deduce that the wooden core of the shield was not particularly thick. The fact that the shield was so narrow would also play a part in its lightness.

Since *thureophoroi* were light infantry, they had to be able to operate in rough terrain, so the lightness of the shield was very important. A terracotta figure of two men competing in *thureomachia* shows one figure lifting the shield high above his head to defend against a sword stroke from his opponent, demonstrating the lightness of the shield.[64] The wooden core of the shield would in all likelihood have been covered with ox-hide, giving the wood an extra layer of protection from weaponry, as well as from the natural elements as wood tends to warp through exposure.[65] The shield is clearly depicted with a spined boss, with the spine or midrib running vertically down the shield, and would have had a central horizontal hand grip which was protected by the boss.

5.3.2 Dóru Spear

Plutarch writes that the *thureophoroi* were armed with the *dóru*, a generic Greek term for spear. The *dóru* spear of the *thureophoroi* is described as being 'much shorter' than the Macedonian *sarissa*. The iconographical evidence shows numerous *thureophoroi* armed with spears. Most appear to be a little taller than their bearers, approximately 2 metres in length. Plutarch's description states that the *thureophoroi* 'were effective in fighting at a long distance, because they were so lightly armed, but when they came to close quarters with the enemy they were at a disadvantage'. From 'effective in fighting at a long distance' we can deduce that they were range combatants armed with a throwing spear or javelins. The length and girth of the spears in the iconographical sources depicting *thureophoroi*, and the fact that Plutarch calls their weapon a spear (*dóru*) as opposed to a javelin (*akontion*), suggests that the weapon was the dual-purpose *longchē*, which could be used for both thrusting and throwing.[66] The *longchē* gave the *thureophoroi* the obvious ability to engage in hand-to-hand combat and yet fight 'at a long distance' using the *longchē* as a missile.

Another possibility could be that the *thureophoroi* were armed with a thrusting spear for close combat but also carried several javelins for range combat. However, the majority of depictions of *thureophoroi* show them carrying only one spear, and it is difficult to say whether they were armed with javelins and a thrusting spear or simply carried two or three *longchai*.

It is interesting to note that the tombstone of Dionysios the Bithynian shows him armed with a long spear and followed by his attendant, who carries Dionysios' *thureos* shield and two more spears or perhaps javelins.[67] It is possible that attendants like that of Dionysios acted as a support runner on the battlefield, running back and forth, providing the warrior with spare javelins or spears. The tombstone also shows that mercenary work under the right paymaster, such as the immensely wealthy Ptolemaic and Seleucid empires, could be lucrative if Dionysios was able to afford an attendant.

5.3.3 Sword

Several depictions of *thureophoroi* show them carrying a type of sword as a secondary weapon.[68] The most iconic of these is perhaps the Stele of Dioskourides, which depicts the warrior as a *thureophoroi* mercenary.[69] Dioskourides is shown in a fighting stance, bearing a white *thureos* shield in his left hand and in his right hand holding a short sword raised above his head, ready to strike. The sword has a bilobate pommel, a popular design in the Hellenistic era, and a straight blade which quickly tapers down to a sharp point, making it ideal for thrusting. The type of sword is well attested among mercenary soldiers of the time.[70]

The small terracotta statuette depicting two youths competing in the *thureomachia* clearly show both bearing *thureoi* shields and wielding straight-bladed short swords for their duel. The fact that they wielded swords during their competition shows the growing importance during the period of the sword as a weapon, it clearly being an important part of the *thureophoroi*'s armament.

5.3.4 Armour

Thureophoroi are shown without body armour, wearing only simple short-sleeve tunics. As light troops, primarily acting as skirmishers who were 'effective in fighting at a long distance', it made sense that they carried very little armour. *Thureophoroi* are, however, almost always depicted wearing helmets. The majority of the helmets on the Sidon Stelae are classified by Sekunda as Sidon-Type A, which was a simple crested, domed helmet, with a visor-like brim, but other varieties are shown as well.[71] The tombstone of Eubolos from Tanagra (275–250 BC) displays two *thureoi* and a Boeotian helm on the pediment, and we can assume that Eubolos fought as a *thureophoros* and wore a Boeotian helm as head protection.

5.3.5 The Thōrakitai

There are some depictions of *thureophoroi* clad in body armour, and scholars have interpreted these to be the *thōrakitai* (θωρακίται), which translates as 'cuirassiers'. There are several images of soldier carrying *thureoi* shields and spears and wearing *linothōrax* armour on a mural from Alexandria.[72] There is also the Sidon Stelae displaying a *thureophoros* in armour, which is the 'Stele of Salamas, son of Moles of Adada' (**see plate 28**).[73] Salamas is armed with a thrusting spear, carries a *thureos* shield and is clad in a chain mail vest, yet another possible introduction to the Hellenistic world by the Gallic invasions of the 270s.

The *thōrakitai* are described in literature as supporting other light infantry, fighting in rough terrain, protecting the flank of the heavy infantry of the phalanx and protecting the baggage train and marching column.[74] The description of their roles in combat and on campaign clearly places them in the category of light troops. The *thōrakitai* can thus be interpreted as heavier versions of the *thureophoroi*, yet still be classified as light infantry based on their function and tactical use.[75]

The literature is silent on the arms carried by the *thōrakitai*, but those who are depicted in the Alexandrian mural and the funeral Stele of Salamas shows them armed in the same manner as the standard *thureophoroi*. The only difference seems to be that the *thōrakitai* wore some form of body armour. The *linothōrax* worn by the *thōrakitai* in the Alexandrian mural has already been discussed (see Chapter 2.1.3, Defensive Equipment – Body Armour). The armour was flexible and light and did not restrict the movement of its bearer, and thus served as ideal armour for light infantry.

As mentioned, Salamas, as a *thōrakitai*, wore chain mail. This type of armour consists of interwoven metal rings, all linked together, which allows the vest to take the shape of the wearer, giving him immense freedom of movement in comparison to other armour types (**see plate 29**). The weight of the vest on the shoulders was lessened immensely through the assistance of a waist belt, which carried part of the weight. A type of padded armour, not unlike the medieval gambeson, was worn underneath to help absorb blows and prevent the metal rings from chafing. This quality of lightness in comparison with other metal cuirasses, and the dexterity it allowed, made it a very fitting type of armour for light infantry.

5.4 Thureophoroi and Thōrakitai at War

From iconographical evidence and the description of their function in battle and on campaign in literature, it is clear that the *thureophoroi* acted as light infantry. They were dual-purpose infantry who could act as range infantry, using javelins or throwing spears, and also engage in melee combat with spears or swords. Their oblong shields gave them sufficient cover, yet remained light enough not to hinder their movement.

Pitched Battle

In many ways, the *thureophoroi* can be seen as a link between the *psiloi* range units and the heavy infantry of the phalanx. In major pitched battles, they could fight in the main battleline, being deployed on the vulnerable flanks of the phalanx, protecting them from flanking manoeuvres. Molon, for instance, placed his *thureophoroi* between the heavy infantry and the cavalry in his final battle against Antiochus the Great.[76] At the climactic Battle of Raphia (217 BC), Antiochus placed the Greek mercenary *thureophoroi* on the right flank of his phalanx.[77] Philopoemen, at the Battle of Mantinea (207 BC), made use of the heavier version of the *thureophoroi*, the *thōrakitai*, to protect the flanks of the Achaean phalanx.[78] These *thōrakitai* also acted to support other light troops during a vanguard attack before regrouping to form on the flanks of the phalanx, emphasizing their diverse roles in combat.[79]

Under the right circumstances, *thureophoroi* could take the offensive and even beat back heavy phalanx units. At Raphia, the Ptolemaic elite guard phalanx had its formation thrown into disarray by an elephant change. The mercenary *thureophoroi* of Antiochus capitalized on the opportunity and charged the phalanx, driving it from the battlefield.[80] Similarly, the Aetolian mercenary *thureophoroi* in Ptolemaic employ at the Battle of Panion (200 BC) were able to push back a Seleucid phalanx of heavy infantry. Polybius quotes Zeno of Rhodes, writing that 'the phalanx, outmatched in agility and forced backwards by the Aetolians, retired step by step'.[81] Since the phalanx was 'outmatched in agility', we may assume that the mercenaries rushed in, pelting the phalanx with their spears and then retiring before the heavy infantry could close with them. The constant bombardment would have disrupted the formation of the phalanx. Because of the speed and agility of the *thureophoroi*, it is possible that they were using their mobility to move around and outflank the Seleucid phalanx, threatening the vulnerable sides, forcing the phalanx to withdraw since it was less able to manoeuvre.

Broken Terrain

Thureophoroi, because of their light equipment, were well suited to operate in difficult and broken terrain. The Aetolian *thureophoroi* fighting in the uneven terrain of Mount Tichius at the Battle of Thermopylae (190 BC) were able to beat back 2,000 Roman legionaries.[82] Their shields were light and manoeuvrable, allowing them to move freely in rough terrain, yet large enough to offer defence against missiles and sufficient protection in hand-to-hand combat. The light spears, javelins and swords did little to restrict their movement, allowing them to move through uneven terrain with ease. In the case of the *thōrakitai*, their body armour was light and allowed for good mobility, so they too could operate in difficult terrain.

The *thureophoroi* and *thōrakitai* were also relied upon to protect the vulnerable phalanx and baggage train on the march. When Antiochus the Great moved through the Elbruz Pass, the difficult terrain made it impossible for the phalanx to operate. These heavy troops and the baggage train were thus very vulnerable while moving through the pass from attacks by lighter infantry. Antiochus therefore placed the *thureophoroi* and *thōrakitai* to defend the rear and flanks of the marching column, while his Cretans acted as the vanguard.[83]

5.5 Conclusion on the Thureophoroi and Thōrakitai

The Macedonian phalanx was the dominant military force of the Hellenistic period. Nevertheless, the phalanx was one-dimensional, being severely restricted in its deployment. The *thureophoroi* and *thōrakitai* were the troops that filled the gaps left by the limitations of the Macedonian-style phalanx. These light troops that could operate in rough and difficult terrain, and act as both range skirmishers and melee light troops. Even though they were light troops, their armament allowed them to join the main battleline during pitched battles and provide flank protection for the phalanx. Serving as a link between the standard range light troops and the heavy infantry of the phalanx made the *thureophoroi* and *thōrakitai* an incredibly versatile troop type that was able to deploy under virtually any circumstances, engaging the enemy with either range skirmishing or in a melee.

Roman Velites/Grosphomachoi

As with the Hellenic states in the Eastern Mediterranean, Rome saw the development of its own hybrid infantry who were able to fight as range skirmishers yet also contribute effectively in a melee when the opportunity presented itself, in the form of the Republican *velites*. The Roman *velites*, according to Polybius, were chosen 'from the youngest and poorest' of Rome's citizens, and were 'ordered to carry a sword, javelins and shield'.[84] Being drafted from the 'youngest' of Rome's citizens made them ideal for skirmishing because of their youthful vigour. Their required armament, which was much lighter than that of the heavy infantry, also reflected this. Their armament also fitted with their financial status, being drafted from the 'poorest' of Roman citizens who qualified for military service, which meant that they did not have the financial means to procure the arms and armour of the heavy infantry. One would imagine that being the youngest of those who could be drafted meant that these men had yet to establish themselves in Roman society, so did not yet have the financial means to arm themselves as heavy infantry, lending further emphasis to the link between status and armament.

In a standard Republican legion consisting of 4,500 soldiers, it is believed that the *velites* had a paper strength of 1,200. More than a quarter of the legion thus consisted of *velites*. According to Polybius, the *velites* were not given their own officers, but were divided equally among the various maniples of the heavy infantry of the *triarii*, *principes* and *hastati*. We may thus assume that they were commanded by the centurion of the particular unit to which they were assigned.

There is a common misconception that the inception of the *velites* was in 211 BC during the siege of Capua in the Second Punic War. This is based on a passage from Livy in which an improvised cavalry tactic was used that would allow the Roman cavalry to best the Campanian mounted forces.[85] Livy mentions that the Romans selected the 'young men who by reason of strength and lightness of build were the swiftest'.[86] These were armed with round shields and seven javelins, which are described as similar to those used by the *velites*. These men would then ride into battle with the cavalry, seated behind the rider:

> 'And when they had come within range, at a given signal the *velites* leapt to the ground. Thereupon an infantry line suddenly dashes out from the cavalry at the enemy's horsemen, and while attacking they hurl one javelin after another. By throwing a great number of these against horses and men in all directions, they wounded very many.'[87]

Livy concludes the passage by stating that 'thereafter the Roman side was superior in cavalry also; it was made the practice to have *velites* in the legions'.[88] While the enigmatic statement is confusing, it should not be taken to mean that it was the first time that the *velites* were included in the legion, but rather the first time the tactic was used by the legions. Anders has interpreted the passage to read: 'Thereafter the Roman side was superior in cavalry; it having been established that the *velites* could also be in the legions.'[89]

The idea that 211 BC saw the creation of the *velites* is erroneous for several reasons. Livy himself mentions *velites* fighting at the Battle of the Trebia in 218 BC, some seven years prior to the siege of Capua.[90] Also, within the very passage describing the innovative tactic, Livy says that these light troops were armed with 'javelins with iron-heads like those of the *velites*'.[91] This would imply that the *velites* were already in existence for these new troops to be armed in the same manner as the light troops. Secondly, the term *grosphomachoi* (γροσφομάχος) is the one Polybius uses whenever he refers to the Roman *velites*.[92] The term appears in his writings as early as the Battle of Tunis in 255 BC and is used throughout the Punic Wars up until the defeat of Hannibal at Zama in 202 BC. Polybius, when describing the Roman Army of his day, still uses the term *grosphomachoi* to denote the *velites* of the second century BC.

In the literature, the *velites* fill a variety of roles typically associated with light infantry. We are told by Polybius that the perimeter guard of the Roman encampment was the special duty assigned to the *velites*, being stationed along the entire *vallum* (defensive rampart) and guarding the entrances of the fortified camp.[93] In 211 BC, during the Second Punic War, Hannibal attacked the fortified camp of Appius Claudius, but the *velites* stationed on the *vallum* were able to repel his attacks with their javelins.[94] The *velites* are often deployed to operate alongside cavalry (Polyb. 10.32.2, 11.24.1, 18.19.9, 18.21.1; Livy 31.35.2), used on reconnaissance missions (Polyb. 18.19.9, 18.21.1), and act as a screening force, skirmishing against other light infantry before the main engagement of the heavy infantry (Polyb. 3.73.1, 3.115.1).[95] They were also used to great effect in operating in difficult terrain where the heavy infantry could not. During the Battle of Baecula (208 BC), Hasdrubal's Carthaginians occupied a steep hill:

> 'In the night Hasdrubal withdrew his force to a hill, the summit of which formed a broad table-land. His rear was protected by a river, in front and on either side the hill sloped down precipitously, forming a kind of steep bank, which surrounded the whole position. Below there was another level stretch of ground which also fell away abruptly, and was equally difficult of ascent.'[96]

Scipio, the Roman general in Spain, led the assault on the hill in person with his *velites* and other light troops:

> 'At first their only difficulty was the rough ground over which they were marching, but when they came within range of the infantry stationed on the lower plateau, all kinds of missiles were showered upon them, to which they replied with showers of stones, with which the ground was strewn, and which not only the soldiers but the camp followers who were with them flung at the enemy. Difficult as the climb was, and almost buried as they were beneath stones and javelins and darts, they went steadily on, thanks to their training in escalade and their grim determination.'

When they managed to reach the top of the hill, the *velites* drew their swords and engaged the Carthaginian light infantry in hand-to-hand combat:

> 'As soon as they reached level ground and could plant their feet firmly, their superior mode of fighting told. The light and active enemy, accustomed to fighting and skirmishing at a distance, when he could evade the missiles, was quite incapable of holding his own in a hand-to-hand fight, and he was hurled back with heavy loss on to the main body posted on the higher ground.'[97]

Rome's lightning campaign against the Galatians in 189 BC also demonstrates the ability of the *velites* to operate in difficult terrain (the campaign is discussed in greater detail under 7.3, The Galatian Campaign). The campaign was fought in rough, mountainous terrain where Livy specifically mentions the heavy infantry making slow and difficult progress.[98] Indeed, the entire campaign was fought and won by the *velites* and allied light troops. The campaign serves as another prime example of the *velites* being very capable of fighting in a melee. Livy's account mentions how the *velites* switched from their javelins to their *gladius* swords, fighting and slaying the Galatians who charged down upon them:

> '[S]ome, rushing against the enemy, were wounded from every side, and when they had come to close quarters they were slain by the swords of the skirmishers. This type of soldier carries a three-foot shield and, in his right hand, javelins which he uses at long range; he is also equipped with a Spanish sabre; if he is compelled to fight hand to hand, he shifts his javelins to his left hand and draws his sword.'[99]

In a skirmish between the light troops of Philip V and those of the Roman Republic, we see a similar account of *velites* being more than willing to engage in a melee to great effect:

> 'His [Philip V's] infantry, trained to skirmish in loose order and unprotected by armour, were at the mercy of the *velites* who with their swords and shields were equally prepared for defence and attack. Incapable of sustaining the conflict and trusting solely to their mobility they fled back to their camp.'[100]

The *velites* of the manipular legion of the middle Republic were provided with the opportunity to prove their *veritas*, which was denied to so many of the other light troops of the Graeco-Roman era. The *velites*, Rome's youngest and poorest, went ahead of the legion and were generally the first to engage the enemy in plain sight of their commanders, officers and brothers-in-arms. Polybius tells us that these *velites* often wore a distinguishing headpiece such as a wolfskin over their helms to help their officers (and one would presume their fellow citizen in the heavy infantry) to identify them when they committed acts of bravery or *veritas*:

> 'Each man also wears a headpiece without a crest; which he sometimes covers with a piece of wolf's skin or something of that kind, for the sake both of protection and identification; that the officers of his company may be able to observe whether he shows courage or the reverse on confronting dangers.'[101]

Furthermore, Polybius states that special gifts such as spears and cups were awarded to the men who 'willingly and deliberately' put themselves in danger. These gifts were bestowed upon them in front of the entire assembled legion. Polybius emphasizes that such gifts were only bestowed upon those who fought in skirmishes, not to those who took part in regular battles:

> 'These gifts are not made to men who have wounded or stripped an enemy in a regular battle or at the storming of a city, but to those who during skirmishes or in similar circumstances, where there is no necessity for engaging in single combat, have voluntarily and deliberately thrown themselves into the danger.'[102]

The poorest and youngest of Rome's citizens were thereby given the chance to express the Graeco-Roman world's ideals of *arête*/*veritas*.

The *velites* are last mentioned in the Jugurthine War (112–106 BC), serving under Metellus in 109 BC just before the military reforms of Gaius Marius.[103] It is very likely that the Marian Reforms, which involved doing away with minimum property rights and state-sponsored equipment, meant that those who previously served as *velites* were now bolstering the legions as heavy infantry. The role of light troops was no longer filled by citizens, the place that the *velites* played in war now being filled by the non-citizen auxiliaries.

5.6 Arms and Armour of the Velites/Grosphomachoi

Thanks to Polybius' assessment of the Roman Army of the Middle Republic, we have a fairly good understanding of the arms and armour of the *velites*, perhaps more so than any other light infantry of Rome's long military history. Polybius describes the *velites* as follows:

> 'The youngest soldiers or *velites* are ordered to carry a sword, javelins, and a shield. The shield is strongly made and sufficiently large to afford protection, being circular and measuring three feet in diameter. They also wear a plain helmet, and sometimes cover it with a wolf's skin or something similar both to protect and to act as a distinguishing mark by which their officers can recognize them and judge if they fight pluckily or not. The wooden shaft of the javelin measures about two cubits in length and is about a finger's breadth in thickness; its head is a span long hammered out to such a fine edge that it is necessarily bent by the first impact, and the enemy is unable to return it. If this were not so, the missile would be available for both sides.'[104]

5.6.1 The Grosphus Javelin

Grosphomachoi, the term that Polybius uses for *velites*, translates as 'those who fight with the *grosphus*'. It is interesting to note that the term *grosphomachoi* appears only in Polybius, but the word *grosphus* appears in Strabo and Plutarch.[105] According to Polybius, the wooden shaft of the javelin was 2 cubits (approximately 96cm) in length and its iron head measured about a span (approximately 23cm), giving it an overall length of about 119cm.[106] This corresponds very well with Livy's account of the *velites'* javelin being four Roman feet (120cm) in length.[107] The shaft was fairly thin, measuring about a 'finger's breath in thickness' (1.8cm in diameter). The iron head, measuring a span, was also forged to be very thin so that it would bend 'by the first impact'. The *grosphus* is in all likelihood the same as the *hasta velitaris* attributed to the *velites* by Livy.[108]

According to Lucilius (7.290), the *velites* carried five javelins, but Livy (26.4.4) mentions that up to seven could be carried.[109] This meant that a minimum of four to six javelins were carried in the left hand, along with the shield. It thus becomes evident why the javelin shafts had to be thin ('a finger's breadth') for the *velites* to carry several of them in one hand. The bending of a javelin head, rendering it useless after it had been thrown, gives us some sense of light troops' range combat. Extensive efforts and costs were involved in having the long iron head bend so that it might not be used by the enemy. For the Romans to go to such lengths, it becomes evident that light troops, after expending their own missiles, relied upon the missiles cast by their enemies that missed their mark, to replenish their ammunition. Additionally, such a long and narrow iron head could more easily penetrate deep into the enemy's flesh.

Neither Polybius nor Livy mention that the javelin was cast with the *ankylē/amentum* throwing strap, but Cicero does mention that the *velites* did so.[110] It is highly unlikely that the *velites* did not make use of the *ankylē/amentum*. The benefits and advantages of throwing the javelin by means of a throwing strap would be hard to ignore. The *ankylē/amentum* was well known throughout Italy, and was certainly used by the Etruscans and Samnites. At the Battle of the River Trebia (218 BC), Polybius perhaps even indirectly references the use of the *ankylē/amentum* by the *velites*:

> 'For the Roman javelinmen on foot [*pezakontistai*] had been on hard service ever since daybreak, and had expended most of their weapons in the engagement with the Numidians, while those weapons which were left had become useless from being long wet.'[111]

Polybius states that the *velites* had spent all their javelins and those that remained could not be used, 'being long wet'. Surely the wooden shafts and iron heads of the javelins were not affected by the wet. What could indeed have been affected was the leather *ankylē/amentum* strap. Livy, in his account of the Battle of Magnesia (189 BC), for instance, mentions that these throwing thongs are negatively affected by damp and wet conditions:

> 'A morning mist, which as the day advanced rose up in clouds, spread a general darkness; and the moisture issuing from it, and coming from the southward, wetted everything ... The king's [Antiochus the Great's] soldiers, as the line was so extensive, could not even see their wings from the centre, much less could those at the extremities see one another; and then, the moisture relaxed the strings of their bows, their slings, and the thongs of their javelins.'

Based on this, I would suggest that the *velites* must have used the *ankylē/amentum* for their javelins to have been negatively affected by wet weather at the Trebia (for the advantages of throwing a javelin with an *ankylē/amentum*, see Chapter 8: Javelin & Throwing Strap).

5.6.2 Gladius Hispaniensis

Both Polybius and Livy state that the *velites* carried swords. Polybius uses the term *makhaira*, which is a very generic term for a variety of swords, but Livy refers to the *velites'* sword as the Spanish sword (*gladio Hispaniensi*).[112] According to archaeological finds, the blade measured approximately 60–67 cm, with an overall sword length of 70–80 cm.[113] While Polybius does not mention the exact type of sword carried by the *velites*, he does note that the Romans were armed with the Iberian (Spanish) sword and describes it as good for both cutting and thrusting, since the 'blade is stout and unbending'.[114] Elsewhere, Livy paints a terrifying picture of the horrors the Romans could inflict with this type of sword:

> '[T]hey had seen bodies chopped to pieces by the Spanish sword, arms torn away, shoulders and all, or heads separated from bodies, with the necks completely severed, or vitals laid open, and the other fearful wounds.'[115]

5.6.3 Parma Shield

Both Polybius and Livy state that the *velites* defended themselves with the *parma* (in Greek, πάρμη) shield.[116] Polybius says that the *parma* was 'strongly made, and large enough to protect the man; being round, with a diameter of three feet'. Livy also mentions that the *parma* was 3ft (approximately 90cm) in diameter.[117] This is large for any shield type, and one would assume very heavy for light infantry. It is likely that the *parma* was constructed from a lighter wood than the heavy infantry *scutum* shield. The Kasr-el-Harit *scutum* discovered at Fayuum in Egypt was constructed of three layers of birch wood, while the Doncaster shield is a combination of oak and alder.[118] Although both birch and oak are very strong and durable woods, they are also very heavy. Pliny mentions poplar and willow being suitable for shields.[119] Both these types of wood, while softer and not as durable, are substantially lighter than oak or birch. Since *velites* were not expected to act as heavy line infantry, it was not necessary for their shields to be as strong and durable as that of their heavy infantry counterparts. It is thus also possible that the *parma* consisted of two or even just one layer of wood, as opposed to the three of the heavy infantry. The *parma* is akin in shape and size to the Dark Age Norse or Viking shields. The Norse sagas mention shields constructed of linden wood, which has a similar density to willow and poplar. The *parma*'s weight could further be reduced by not being covered in leather, unlike the infantry *scutum*.

The *parma* was in all likelihood held by a central handgrip with a metal boss, rather than being strapped to the forearm. This would allow for a greater movement range and for the shield to be punched forward and used as a weapon. The handgrip would have been thin to facilitate holding javelins, so that if the *velite* was 'compelled to fight hand to hand, he shifts his javelins to his left hand and draws his sword'. The Mainz pedestal shows a much later Roman auxiliary soldier carrying a shield and two javelins in his left hand, while he has a third javelin in his right. Similarly, the Norse shield's handgrip was thin enough for the warrior Thorgeir (Þorgeirr) in the eleventh-century Icelandic *Fóstbræðra Saga* to grip his shield in his left hand as well as hold his axe, while he used his spear in his right hand to duel against Snorri.[120]

5.6.4 Helmet

As mentioned by Polybius, the *velites* wore plain helmets without any crest. Due to their social status of being the 'youngest and poorest' of Roman society, these helmets must have been of a very simplistic and affordable design. The animal pelt, such as a wolf skin, would have added additional protection by providing extra cushioning to a blow to the head, but the main purpose was to distinguish the warrior from those around him.

5.7 Velites/Grosphomachoi at War

Even though the *velites* filled all the typical roles of light infantry, and were the 'youngest and poorest' of Rome's military force, they were property-owning citizens of the Roman Republic. This, along with the fact that these men were given the opportunity to prove their *veritas* by engaging the enemy in single combat, set the *velites* apart from other light infantry. This meant they did not suffer the same prejudice as their counterparts did.

Fighting Other Light Troops

The *velites* were ideally suited to fighting other light infantry. They were lightly equipped, and being young one would assume they were some of the most energetic and fittest of Rome's fighting force. They were thus nimble and quick enough to skirmish against the enemy, fighting at range with their javelins. Skirmishing at the time was very much a back-and-forth exchange of missiles. During the range skirmish, the large 90cm diameter *parma* shield gave the *velites* sufficient protection against the missiles of the enemy. Once a side had spent their own missiles, they could collect the enemy's missiles that had been thrown at them, and the process would repeat itself. As discussed above, the *grosphus* javelin used by the *velites* was designed to bend upon impact, which denied the enemies a round of missiles to exchange. The *velites* could then collect the enemy's javelins and throw them back without receiving return fire. Alternatively, the *velites* could stand their ground and refuse to allow the enemy light infantry to advance to collect their spent javelins. The *velites*, being armed and trained for melee combat, and more than willing to fight hand-to-hand to demonstrate their *veritas*, could quickly advance and engage the enemy light infantry. Their better gear and training gave them a big advantage over other light troops in a melee. This is reflected in their skirmish against Philip V's Macedonian light infantry, who had been trained to skirmish in loose order and wore no armour, making them vulnerable to the *velites*.

Countering Elephants

The *velites* seem to have been Rome's best counter against elephants. At the Battle of Panoramus (250 BC), the *velites* were used to first harass the enemy elephants as the Carthaginian army tried to cross the river. The Carthaginians drew up their forces with their elephants at the front before the city of Panoramus, which had a deep moat surrounding it. The *velites* continued to attack the elephants by showering them with javelins. Every time the elephants charged, the *velites* retreated to the moat for protection and to resupply with more javelins. Polybius specifically mentions that the armourers in the city were tasked to make an abundant number of javelins for the light troops. The constant harassment by the *velites* eventually forced the elephants to charge the wall. As soon as they reached the wall, they received a hail of arrows and javelins, both from above and from the *velites* in and around the moat. Wounded and panicked, the elephants were driven upon their own ranks and shattered the Carthaginians lines:

> '[S]tudded all over with darts, and wounded past all bearing, they soon got beyond control. They turned and bore down upon their own masters, trampling men to death, and throwing their own lines into utter disorder and confusion.'[121]

At the Battle of the Trebia (218 BC), Livy mentions that the *velites* were specifically used as a counter to Hannibal's elephants:

> 'The *velites*, expressly posted to deal with the beasts, would throw darts at them and make them turn away, and then pursuing them would strike them under the tail, where the skin is softest and it is possible to wound them.'[122]

According to Livy, the *velites* were so effective against the beasts that Hannibal was forced to move his elephants from the centre of the battlefield to his left wing to keep them from panicking and running amok among his own troops.[123]

At Zama (202 BC), Scipio's *velites* were once again used as a screening force against Hannibal's elephants. Polybius tells us that the elephants did much damage to the skirmishers, but the beasts themselves suffered greatly from the javelins of the *velites*.[124] The *velites* were so effective that the elephants were driven off the field, retreating from whence they came or running through the empty gaps between the maniples of the Roman legions.

5.8 Conclusion on the Velites/Grosphomachoi

The *velites* were assigned the greatest variety of tactical roles, such as scouting and reconnaissance, skirmishing, guard duties and operating in difficult terrain. They were Rome's first contact with the enemy and were used to blunt the enemy's attack, being the first to expose themselves to danger and the terrors of battle. They were expected to willingly face these dangers and prove their Roman *veritas*. However, unlike most light infantry, their actions and bravery on the battlefield were recognized and there was a deep appreciation for their courage and risks, which were awarded in front of all their comrades in arms. In many ways, the *velites* embodied Rome's martial values and attitude towards war. The *velites* must have had a powerful psychological effect on the battle. The opening stages of the skirmish would undoubtedly have had a moral impact upon the rest of the clash. The actions of the *velites* would have inspired the heavy infantry before they were to engage the enemy. They would also have had the opposite effect on the enemy, seeing how the *velites* willingly put themselves at risk, fighting both at long range and in melees. They reflected Rome's martial spirit and demonstrated to the enemy what and who they would face in close combat.

Chapter 6

Elite Light Troops of the Greek and Roman World

The Cretan Mercenaries

The most prolific mercenaries of the ancient era were undoubtedly the Cretans. They served as hired soldiers in every part of the Mediterranean world and beyond, with scholars such as Griffith referring to them as 'by far the most prominent Greek soldiers throughout the Hellenistic period'.[1] They were so acclaimed on ancient battlefields that they have been compared with the Swiss mercenaries of the medieval era, and it was not unusual to find Cretans fighting against other Cretans in the same battle and for Cretan leaders to provide mercenaries to both sides of a conflict.[2] Livy, for example, highlights the Roman senate's displeasure during the Third Macedonian War (171–168 BC) when the Cretans admitted to sending their soldiers to serve among the armies of the Roman Republic as well as the Macedonian forces of Perseus.[3] Indeed, the Macedonians were supplied with a greater number of Cretans than the Romans.

Classical Greece

According to Pausanias, Cretans were hired by Sparta to fight in its wars against the Messenians as early as the eighth and seventh centuries BC.[4] While some sources claim that the Cretans took part in the defence of Greece against Persia, Herodotus specifically states that the Cretans did not send aid.[5] More reliable sources such as Thucydides confirm their use as mercenaries as early as the Peloponnesian War. The disastrous Athenian expedition to Sicily, for instance, had a contingent of eighty Cretan archers.[6] Pausanias also mentions Cretan archers honoured on a memorial dedicated to those who died fighting alongside the Athenians during the Peloponnesian War.[7] Thereafter, the Cretans served as mercenaries, particularly as archers, in almost every major war, campaign and battle, from the march of Xenophon's Ten Thousand to Rome's triumph over the Hellenistic kingdoms in the East, and were still extensively used during the last days of the Roman Republic during the civil war that ushered in the era of Imperial Rome.

In 401 BC, Spartan exile Clearchus brought with him a contingent of 200 Cretan archers, along with his 1,000 hoplites and 800 Thracians, in service to Cyrus the Younger in his revolt against his brother, Artazerxes II.[8] After the failed rebellion, the Cretans gave a good account of themselves as archers under Xenophon in his long march back to Greece through hostile territory. Fitting to their reputation, they conducted ambushes and ruses, fought in numerous skirmishes and excelled in operations in difficult terrain.

Sparta consistently made use of Cretan mercenary archers during the Corinthian War (395–387 BC) to supplement its light troops in an attempt to maintain its hegemony over the southern Greek world. At the Battle of Nemea (395 BC), Sparta deployed 300 Cretan archers and gained a victory over the coalition forces of Thebes, Athens, Argos and Corinth.[9] In 388 BC, Cretan mercenary archers were sent by Sparta on a raiding mission to ravage the countryside around Nauplia.[10] In 362 BC, mercenaries from Crete appeared once again when 'Cretan runners' were sent by the Spartan king, Agis, to warn the citizens of Sparta to ready themselves against the approaching army of Epaminondas.[11]

Alexander and the Hellenistic Era

It is no surprise that Cretans served in the armies of Philip II and Alexander the Great. The army which Alexander inherited from his father and crossed the Hellespont into Asia with contained a contingent of Cretan archers. Their exact number is obscured, as they are grouped in the sources with the Agrianians and the Macedonian contingent of archers and are given a combined number of 1,000 soldiers.[12] Throughout Alexander's campaigns, the archers and Agrianians are more often than not paired to fight together. Gaebel highlights that the archers were grouped fifteen times with the Agrianians in the army of Alexander.[13] These pairings saw some of the toughest fighting throughout Alexander's campaigns. They were often taken on forced marches, operated in difficult terrain and on numerous occasions are specifically mentioned as chosen by the king himself to go on special missions.

The Hellenistic era, which was ushered in by the death of Alexander and the break-up of his empire, saw a significant increase of Cretan mercenaries throughout the Mediterranean world. The wars fought for dominance by his Successors (*Diadochi*) and the various conflicts which saw the eventual rise of Rome made sure there was no shortage of mercenary positions for Cretans to fill.

Antigonids

Cretans served extensively under the Antigonid dynasty which ruled Macedonia. Demetrius Poliocretes employed Cretan archers for his epic siege of Rhodes

(304–305 BC).[14] *Symmachoi* treaties between Antigonus Doson and the Cretan city-states of Hierapytna and Eleuthernae highlight the king's attempt to establish a stable supply of Cretan mercenaries for his wars. They also provide us with valuable information on daily pay (at one Alexander drachma per day in the case of those recruited from Hierapytna) and the number of soldiers these treaties demanded.[15] Doson's nephew and successor, Philip V, fielded numerous Cretans in his campaigns during the Social War (220–217 BC). A number of these were sent in demand of a treaty between Philip and the city-state of Polyrrhenia, not unlike that between Antigonus and Hierapytna.[16] At the same time, the Conossians sent 1,000 Cretans to aid the Aetolians to fight against Philip. Polybius specifically states that the above-mentioned Cretans from Polyrrhenia and Conossias served on their respective sides for the remainder of the Social War.[17] Furthermore, Philip's garrison troops at Corinth included 800 Cretans.[18] In Philip's war with Rome, he also made extensive use of Cretans. A generation later, Philip's son, Perseus, took on Rome once again during the Third Macedonian War (171–168 BC). When the king assembled his entire army, it included nearly 3,000 Cretans.[19]

Seleucids and Ptolemies

Cretans served extensively in the many conflicts between the Seleucids and Ptolemies known as the Syrian Wars, which were fought for dominion over the area of Coele-Syria. The Battle of Raphia (217 BC), which was one of the largest conflicts of the Hellenistic era, saw the Seleucids field 1,500 Cretans and a thousand Neocretans (see A Note on Neocretans below), while the Ptolemies fielded 2,000 Cretans and 1,000 Neocretans.[20] Cretans were present in almost every Seleucid campaign up to the Battle of Magnesia (189 BC). Antiochus the Great, for instance, employed Cretan *symmachoi* against the rebel Molon and numerous Cretans accompanied the king on his great *anabasis* (212–205 BC) to reconquer the breakaway eastern provinces of Bactria and Parthia.[21] They were also present in his climactic final battle against Rome at Magnesia in 189 BC.[22] After the defeat at Magnesia, the Treaty of Apameia prohibited the Seleucids from recruiting west of the Taurus Mountains, which meant the Seleucids were no longer allowed to hire mercenaries from Crete. Cretans were notably absent from the Daphne Parade of Antiochus IV Epiphanes in 166 BC.[23] They do, however, appear once again in the armies of Demetrius II when he arrived in Syria in 147 BC with a mercenary force built around a core of Cretans to retake the Seleucid throne.[24] In 162 BC, during the dynastic struggle between Ptolemy VI and Ptolemy VIII Physcon, the latter hired a large contingent of Cretans in an attempt to overthrow his brother.[25]

Hellenistic Sparta

Throughout the Hellenistic period, Cretans appear as mercenaries in the armies of numerous smaller states. Sparta employed Cretans in the defence of their city against Pyrrhus of Epirus in 272 BC.[26] Cretans also served as mercenaries in the armies of the Spartan reformer Cleomenes in his wars against the Achaeans and Macedon.[27] A thousand of these even accompanied him on his exile to Egypt.[28] During the Social War, Sparta hired its own Cretans to oppose the forces of Philip, which demonstrates just how politically fractured the island of Crete was.[29] Nabis (ruled 207–192 BC), the last of Sparta's socialist tyrants, also employed Cretans and fielded them against the Achaeans, the might of Rome and even his own people who opposed his rule. Before his war with Rome, Nabis even provided aid to the Republic by sending 600 Cretans to serve during its wars in Greece. When hostilities broke out between Rome and Sparta, Nabis had as many as 2,000 Cretan mercenaries at his service.[30] Prior to the war with Nabis, the Republic had offered a peace treaty to the tyrant, which he refused. The treaty was very unfavourable for Nabis, explicitly mentioning that Sparta was not allowed to hold any city in Crete or make alliances with Cretan city-states. This clearly highlights that Rome and its allies recognized that Nabis' power was based on his mercenary forces, Cretans in particular.

Achaean League

The Achaean League also used Cretans to bolster their forces. The army with which the Achaeans took on Nabis in 192 BC and Philopoemen's Messenian expedition in 183 BC both contained Cretan mercenaries among the Achaean forces.[31] Cretan mercenaries must have had a profound impact upon the Achaean army, because by 171 BC, at the start of the Third Macedonian War, the Achaeans sent aid to the Romans in the form of 1,500 men 'mostly armed in the Cretan fashion'.[32]

Pergamum

Cretan mercenaries also appeared in the armies of the Kingdom of Pergamum in Asia Minor. In 208 BC, King Attalus had with him a contingent of Cretan mercenaries in his war with Philip V. There was also a *symmachoi* treaty between the Cretan city-state of Aptara and King Attalus which gave the king special recruiting rights in Crete.[33] Eumenes II fought alongside Rome at the Battle of Magnesia and fielded 500 Cretans, who played an important part in the clash against the Seleucids.[34] Eumenes also provided assistance to Rome by sending Cretans and other light troops during the Galatian War (189 BC).[35] There is further evidence of treaties of alliance between Eumenes II and several Cretan cities, which might have been *symmachoi* treaties, but the sources are very fragmented and inconclusive.[36]

Rome

During the Second Punic War (218–201 BC), Rome was sent 500 Cretan archers and 1,000 *peltasts* by its ally, Hiero of Syracuse.[37] Livy mentions that these Cretan bowmen and *peltasts* were specifically provided because they were 'capable of acting against the Balearics and Moors and other tribes who fight with missile weapons'.[38] In 197 BC, in preparation for its war with Philip V, Rome had supplemented its own troops with allied contingents from Greece, which included 500 Cretans.[39] As with the Achaeans, the Cretans had a noted impact on the fighting styles of other Greek cities; we see that along with the 500 Cretans, the Apollonians sent 300 soldiers to aid Rome who were 'armed like the Cretans'.[40] I have already mentioned that Rome used numerous Cretan mercenaries during the Third Macedonian War against Perseus. Cretan mercenary archers were even used as part of the senatorial forces in 121 BC to suppress the Roman mob, instigated by Gaius Gracchus.[41] Caesar himself also employed Cretan archers in his Gallic Wars, while Pompey had Cretan archers among his auxiliaries during the Civil War with Caesar.[42]

Why Cretans?

From the above, it is clear that Crete became an important reservoir of mercenary manpower, especially during the Hellenistic Period, for the various warring states across the Graeco-Roman world. Cretans were clearly highly sought-after mercenaries. But why were Cretans in such demand, and why were they so readily available in such high numbers to be hired as soldiers of fortune?

Along with Sparta, Crete's military reputation was very well established throughout the Mediterranean. The islanders' martial reputation is reflected when Cleomenes III of Sparta looked upon the Ptolemaic army with disdain in 222 BC. The reason for his scorn was because despite the large numbers of Syrians and Carians in the Ptolemaic army, Cleomenes brought with him 'real soldiers' from the Peloponnese and Crete. He addressed a Ptolemaic dignitary:

> '"Don't you see," he said, "that nearly three thousand of them are from the Peloponnese and about a thousand are Cretans, and I need but make a sign to these men and they will all put themselves joyfully at your service. Once they are united whom have you to fear? The soldiers from Syria and Caria I suppose!"'[43]

The Cretan *poleis* were dominated by an aristocratic class of citizen-warriors who ruled over non-citizen serfs and slaves. A martial spirit was thus endemic to Cretan society, and legend has it that Lycurgus modelled his Spartan military regime and the *Agoge* military education of Sparta's youth on that of Crete.[44] Strabo also mentions that the Cretan *Andreia* (common meals or a system of messes similar to those of Sparta) were to promote a sense of martial spirit among Crete's youth:

> '[F]rom boyhood they should grow up accustomed to arms and toils, so as to scorn heat, cold, marches over rugged and steep roads, and blows received in gymnasiums or regular battles; and that they should practise, not only archery, but also the war-dance.'[45]

Indeed, citizenship of the many independent city-states of the island depended on military training and participation in the above-mentioned common meals. Chaniotis, in his *War in The Hellenistic World: A Social and Cultural History*, highlights that 'war was the most important feature of Cretan everyday life (and the frequent cause of death)'.[46] Warfare and violence were very much a part of the island and born out of territorial pressure due to the fifty to sixty independent city-states which dotted the restricted landscape. Cretan society focused on a subsistence economy based on farming. The land was owned by the small portion of citizen-warriors and cultivated by the dependent population of serfs, who paid tribute to the privileged elite warrior class. The system was in itself very unstable and there was a variety of factors that could disrupt the delicate balance. Chaniotis mentions but a few, such as bad harvests, climate change, destruction and loss of cultivated land due to the continuous warfare on the island, escaped slaves, serf uprisings and the concentration of landed property in a few large estates.[47] When the equilibrium of the system was disrupted, it resulted in food shortages and numerous citizens without land. This in turn led to wars of conquest of neighbouring territories, which only exacerbated the situation across the island. These landless citizens, who were preoccupied with military training and a warrior ethos hankering back to that of Homeric times, naturally saw mercenary service as a means of survival. Large-scale migrations to Asia Minor and Egypt, where Cretan mercenaries settled as entire communities (*Kreton politeuma*), are well attested. These large numbers of Cretans serving as mercenaries during the period are thus well explained by the various social-economic factors which impacted the island and allowed for numerous Cretans to be hired to fight.

At times, the recruitment of Cretans as mercenaries was facilitated by allied or *symmachoi* treaties, which allowed for special recruiting rights between the

Cretan *polis* and the mercenary employer. A wonderful example of such a treaty between Rhodes and the Cretan city of Hierapytnia has survived, stating the clauses of the contract in great detail, including the daily pay for the mercenaries.[48] The Cretans who fought for Antiochus the Great against the rebel Molon are described by Polybius as '*symmachoi*'.[49]

These islanders also provided their employers with a sought-after specialised fighting style as archers. Archery as a martial skill was not common among the Greeks or Romans, and Cretans thus had a monopoly when it came to archery. Cretan mercenaries were synonymous with archers, so much so that Diodorus states that Apollo himself taught the islanders how to use the bow:

> '[A]nd as the discoverer of the bow he [Apollo] taught the people of the land [Crete] all about the use of the bow, this being the reason why the art of archery is especially cultivated by the Cretans and the bow is called "Cretan".'[50]

From Strabo's comment above, we can see that the island's constitution and raising its youth was intertwined with archery, as they grew up 'accustomed to arms' and practised 'not only archery, but also the war-dance'.

According to Plato's *Laws*, it was the island's geography and its very landscape that forced Crete's martial citizens to adopt the bow as their primary weapon:

> 'As you may notice, Crete, as a whole, is not a level country, like Thessaly: consequently, whereas the Thessalians mostly go on horseback, we Cretans are runners, since this land of ours is rugged and more suitable for the practice of foot-running. Under these conditions we are obliged to have light armour for running and to avoid heavy equipment; so bows and arrows are adopted as suitable because of their lightness. Thus all these customs of ours are adapted for war.'[51]

In the literature, Cretans are almost always described as archers. If not said to be archers, they are simply referred to as 'Cretans', and even in these cases they are often regarded as bowmen. While they are not associated with any other weapon, it is unlikely that they only trained in archery and carried no other arms. It is, however, safe to say that the bow was the primary weapon among the Cretan mercenaries and the reason why they were so highly sought after.

Cretan Infamy and the Art of Deception

Cretans were the most famous, and in many cases the most infamous, mercenary forces of the Graeco-Roman world. Throughout history, mercenary service has often been thought of as a less-than-savoury career. This is understandable, as mercenaries are essentially hired killers who are first and foremost loyal to their coin before any moral convictions, political obligation or their paymaster's cause. According to Polybius, the Cretans were the worst of all peoples when it came to the love of money, which they placed above all else:

> 'Their [Cretans'] laws go as far as possible in letting them acquire land to the extent of their power, as the saying is, and money is held in such high honour among them that its acquisition is not only regarded as necessary, but as most honourable. So much in fact do sordid love of gain and lust for wealth prevail among them, that the Cretans are the only people in the world in whose eyes no gain is disgraceful.'[52]

We have also seen that it was not strange for Cretan mercenaries from the same city-state to be fighting on opposite sides of the same battle. During the Peloponnesian War, Thucydides highlighted that 'far from coming to assist their own colonists, they were prepared to attack them for pay'.[53]

Cretans were used by tyrants to enforce their rule and suppress citizens and local populations who resisted their despotic regimes. Nabis, whose power base consisted of large numbers of Cretan mercenaries, used them as assassins to hunt down the Spartan citizens who opposed his rule:

> 'Finally, in the towns, renting through unsuspected agents the houses next door to those in which the exiles resided, he [Nabis] introduced Cretans into them, who breaking down the walls and shooting through the existing windows slew the exiles in their own houses either when standing or reposing, so that for the unhappy Spartans there was no place to fly to and no moment at which their lives were safe.'[54]

Similarly, the Seleucid king Demetrius II, whose power was based on his large pool of Cretan mercenaries, used them as a tool for oppression, a period known as the 'Cretan Tyranny'. The leader of the Cretans and other mercenaries, Lasthenes (a Cretan himself), managed to gain high political office and immense influence over the king, so much so that in a letter to Jonathan, the Jewish high priest, Demetrius II referred to Lasthenes as 'father' and 'kinsman'. These are honorary titles as opposed to actual family ties, but they do reflect the influence Lasthenes carried. Lasthenes abused his new position of power, and in the words of Griffith 'reduced the army from a weapon of war to an instrument of oppression', extorting the local population and plundering the kingdom's resources to pay his mercenaries.[55]

Demetrius had disbanded the citizen army of the city of Antioch while retaining the mercenary army, which caused great discontent among the regular troops. The oppressive treatment by Lasthenes and the disbanding of the army led to what is known as the 'Antiochian Riots', which spilled over into a full-on assault of the royal palace. In response, the Cretan mercenaries, supported by 3,000 Jewish mercenaries, took up position on the palace roof and from there loosed their arrows into the mob, and 'because they were so remote from them [the rioters] by their height, that they suffered nothing on their side, but did great execution on the others, as fighting from such an elevation'.[56] One Jewish source claims that the Jewish mercenaries alone killed 100,000 citizens of Antioch. This is obviously an exaggeration, but does give some idea of the extent of the massacre.

Clearly, the right amount of coin and the lack of social ties meant that these mercenaries could be used by tyrants to do their underhanded dirty work. Diodorus paints a vivid picture of Demetrius' mercenaries as an extension of his will:

> 'He got together a company of mercenary soldiers against them, and commanded that the citizens should be disarmed. Since the Antiochians refused to hand over their weapons, he killed some of them as they fell into his hands, and others he murdered in their own houses, together with their wives and children.'[57]

Polybius also tells us that Cretans excelled in all military matters that concerned deception:

> '[B]oth by land and sea [they] are irresistible in ambuscades, forays, tricks played on the enemy, night attacks, and all petty operations which require fraud, but they are cowardly and down-hearted in the massed face-to-face charge of open battle.'[58]

Antiochus the Great's war against his uncle and would-be usurper, Achaeus, serves as an example of the intrigue at which Cretans excelled. Achaeus was besieged in the city of Sardis by Antiochus' Seleucid forces. Polybius writes that the city was impossible to take through an all-out assault due to its strong fortifications, so Antiochus begrudgingly settled in to starve the city into submission. However, one of Antiochus' Cretans, an officer named Lagoras, whom Polybius describes as a man 'with considerable military experience', went to great efforts to reconnoitre the city. He found an unlikely place that was regarded by the city's garrison to be too precipitous to scale. Lagoras, however, devised a plan for a daring clandestine night operation, sneaking into the city by means of scaling the difficult wall and opening the gates for the main army to enter. Lagoras' plan succeeded and the city was captured through the night attack.[59]

Cretan intrigue in the story of Sardis does not end there. Achaeus fell back to the citadel, with little hope of escaping. The Ptolemies, who were in alliance with Achaeus, had hoped to rescue their trapped ally. Sosibius, the governor of Ptolemaic Egypt, approached one of his Cretan officers named Bolis, who was 'regarded as a man possessed of superior intelligence, exceptional courage, and much military experience'. Bolis was given the task of rescuing the trapped Achaeus and bringing him to safety. Bolis was, however, a friend and relative of Cambylus, a Cretan who was captain of Antiochus' Cretan mercenaries. The two Cretans met in secret to discuss the matter and decided to double-cross Sosibius and Achaeus. Polybius highlights the deceptive nature of what he considered as actions highly typical of Cretans:

> '[T]hey discussed the matter from a thoroughly Cretan point of view. For they did not take into consideration either the rescue of the man in danger or their loyalty to those who had charged them with the task, but only their personal security and advantage. Both of them, then, Cretans as they were, soon arrived at the same decision, which was to divide between them in equal shares the ten talents advanced by Sosibius and then to reveal the project to Antiochus and undertake, if assisted by him, to deliver Achaeus into his hands on receiving a sum of money down and the promise of a reward in the future adequate to the importance of the enterprise.'[60]

Bolis then lured Achaeus out of the citadel under the pretence of smuggling him and a small contingent of his followers to safety. Achaeus attempted to disguise himself so that Bolis would not know which of the group of men was Achaeus, but Polybius tells us that Achaeus was 'trying to play the Cretan with a Cretan', and upon recognizing Achaeus, Bolis seized him and he was handed over to Antiochus.[61]

A Note on Neocretans

As we have seen, Neocretans appeared alongside Cretans in the armies of the Ptolemies, the Seleucids and Aetolia.[62] The term Neocretan and how it differentiated from Cretan has been a debated topic with a variety of theories. Walbank and Griffith support Tarn's suggestion that the Neocretans were Cretan citizens armed in a different manner than other Cretans.[63] Launey built upon this by suggesting that the Cretans were traditionally archers, while the Neocretans were light infantry who fought hand-to-hand and were armed with small *peltai* shields.[64] Head and particularly Sekunda argue against this, proposing that the Neocretans were simply newly recruited Cretans from youth associations or *neoi*, typical of Cretan society, and that '*neoi Kretes*' transformed into Neocretans.[65] Spyridakis and Yalichev suggest that the Neocretans were second-class citizens from Crete, similar to the Spartan *neodamodeis*.[66] Others such as Taylor have suggested that Neocretan was a pseudo-ethnic term, similar to Hellenistic *Tarentines*, and they were not ethnically Cretan, but were armed and fought in the Cretan style.[67]

None of these theories can be supported by clear evidence, however, and the identity of the Neocretans remains a mystery. Launey's proposition that the Cretans were archers and the Neocretans light melee troops with shields is problematic. Head points out that there is no evidence to suggest that the Neocretans carried shields, while the Cretan mercenaries serving with Xenophon and Perseus are described as carrying shields but are identified as Cretans, not Neocretans.[68] Similarly, the Cretans of Antiochus carried small shields but were not identified as Neocretans.[69] The best evidence we have is that the Cretans and Neocretans in the Seleucid army at Raphia fought side by side. While Polybius differentiates between them when describing the forces of Antiochus, he groups them together in one unit when he details the Seleucid deployment.[70] From this we have to assume that the Cretans and Neocretans fought in a similar manner, ergo their arms and armour could not have differed all that much.

6.2 Cretan Arms and Armour

6.2.1 The Bow

As discussed, the primary weapon of the Cretans was the bow (**see plates 1–3**). There are very few references in literature to Cretan bows; even less so when it comes to iconography. We have already seen that Greek archers from the mainland made use of the composite bow, and it is thus very unlikely that archers of such elite status as the Cretans would have used a simple self-bow. There is at least one account from the third-century BC poet Callimachos that confirms the composite nature of the Cretan bow, as he mentions a Cretan mercenary dedicating his 'horn bow' to the god Sarapis.[71] Xenophon is the only historical source that discusses the Cretan bow, and even he only touches briefly upon it. When reading the *Anabasis*, it appears at first glance that the Cretan bow was weaker than those used by the Persians, as the Persian archers outranged the Cretans – 'for the Cretan bowmen not only had a shorter range than the Persians'.[72] This is, however, easily explained by the fact that the Cretans used heavier arrowheads than the Persians. This gave them better hitting power, but in turn the heavier weight would reduce the arrow range. This is confirmed later on by Xenophon when the Cretans started using Persian arrows and practised shooting a longer range:

> 'The Persian bows are also large (like those of the Cretans), and consequently the Cretans could make good use of all the arrows that fell into their hands; in fact, they were continually using the enemy's arrows, and practised themselves in long-range work by shooting them into the air.'[73]

It is also important to note that Xenophon describes the Cretan bows as 'large', like those of the Persians. The shooting of the same arrows is interesting too. Any archer will tell you that the weight of an arrow and the spine of the shaft (how much flex is in the shaft) is dependent on the strength or draw-weight of a bow. Shooting arrows that are under-spined or underweight will damage the limbs of the bow, while those that are too heavy or over-spined will not fly true or very far.[74] For the Cretans to use Persian arrows with their bows would suggest that their bows were not only similar in size, but also in draw-weight or power.

6.2.2 Shields

The Cretans in Xenophon's *Anabasis* are described as carrying bronze *peltai* shields.[75] Polybius' account of Antiochus' 2,000 Cretans at Mount Labus also describes the Cretans as 'shield bearing', or '*aspidiōtai*'.[76] Plutarch specifically mentions Cretan *peltai* shields and quivers among the arms captured from the Macedonian army of Perseus.[77] As mentioned, visual representations of Cretans are exceptionally rare. There are only two certain depictions of Cretan archers in iconography that I am aware of. Both are funerary stele from Demetrias. The first is the Stele of Chaironides of Lyttos, which dates from the end of the third century BC.[78] Chaironides is depicted wearing a white tunic, black cloak and bronze helmet. In his right hand he carries a bow. He was wealthy enough to have a servant, who carries his *peltē* shield and a set of javelins. The second depiction is the Stele of Thersagoras from the city of Polyrrhenia, which dates from the end of the third century to the beginning of the second century BC.[79] The Stele of Thersagoras is unfortunately less well preserved, but it does appear as if the figure has a shield slung across his back and is carrying a bow in his left hand.[80] The iconography thus supports the literature that Cretan archers made use of *peltai* shields.

The *peltai* used by the archers were in all likelihood not too different from those used by the Macedonian phalangites (**see plate 25**). As mentioned, the smaller *peltē* shield could be strapped to the forearm, leaving both hands free to use the bow. For experimental archaeological purposes, I constructed a *peltē* with a wooden core and a thin bronze covering, based on the *peltē* dimensions given by Asclepiodotus.[81] The method of attaching the shield is very similar to the *peltē* displayed on the kylix in Basel, which shows the inside of the archer's shield, and the strapping of the *peltē* depicted in a mural from the House of Menander.[82] I found that the *peltē* and its particular method of attaching the shield to the forearm allowed me to use the bow and shield simultaneously with ease, with very little impact upon my shooting capabilities (**see plate 13 and 30**).

The numerous depictions of kneeling archers have already been highlighted.[83] Dropping down on one knee and crouching down makes you a smaller target on the battlefield. I have tried this shooting position on several occasions and found it a very comfortable one to shoot from as it gives an extra element of stability. Adding the shield to the equation while the archer took this kneeling position means the majority of the body was hidden behind the shield, giving a greater amount of protection to the archer (**see plate 30**).

6.2.3 Body Armour

We have seen that some archers are depicted in light forms of body armour such as the *linothōrax*. It is possible that some Cretans were armoured in a similar fashion, but neither Chaironides nor Thersagoras are shown as wearing body armour. The Cretans in Xenophon's *Anabasis* are also mentioned to be unarmoured.[84] Both Chaironides and Thersagoras, however, do have helmets. Thersagoras' helm is not easy to make out due to the ravages of time, but Chaironides is shown with a small open-faced bronze helmet.[85] An open-faced helmet would provide decent protection to the archer, yet would not obscure his vison. Greaves are not mentioned in the literature and not displayed in the iconographical evidence. This is likely due to Cretans relying on being as agile as possible and to allow them to operate in difficult terrain, which was their speciality.

6.2.4 Other Equipment

While the bow was the main weapon of the Cretans, it has already been mentioned that other weapons were not ignored. Numerous amounts of sling bullets which were excavated across the island testify to the popularity of the sling among the Cretans.[86] Chaironides' servant, who carries his shield, also carries a set of javelins for his master and we may assume that Chaironides would alternate between bow and javelin, depending on the need. It is also interesting to note that a fragment of a relief from Perinthus (first century BC) displays a weapon set which is akin to that of Chaironides.[87] A small round shield lies on top of a bow, quiver and a loose javelin. A curved dagger is also on display just below the shield. It is more than likely that these could have belonged to a Cretan mercenary. As with many other light infantry, it is very possible that Cretans would have been armed with a secondary weapon such as a sword or dagger for hand-to-hand combat. This is perhaps even hinted at in literature, as Polybius' account of the siege of Psophis (219 BC) seems to have the Cretans engage their enemies in hand-to-hand combat after breaking their formation with volleys of arrows:

> '[T]he Cretans, attacking the mercenaries who had sallied from the upper gate, forced them to fly in disorder, throwing away their shields. Pressing close on their heels and cutting them down, they entered the gate together with them, and thus the city was taken from every side at once.'[88]

6.3 Cretans at War

Close-Range Snipers

As with other Greek archers, Cretans primarily shot their arrows at close range, preferring hitting power over range. It was only when the Cretans in Xenophon's army started shooting the lighter Persian arrows that they 'practised themselves in long-range work by shooting them into the air'. The range practice of 'shooting them into the air' obviously refers to the firing of overhead parabolic shots. Cretans fought in the thick of battle, acting as snipers, picking out individual opponents as opposed to shooting large volleys of arrows into groups of enemies.

As an example, the Cretans of Philip V, at the siege of Psophis, attacked the Elean mercenaries who sallied out for a counterattack. The Cretans drove back the attack with their arrows, which caused the mercenaries to discard their shields and flee in panic. The Cretans then pursued, and were so close upon their heels that they were able to cut down many of them and rushed through and seized the open gates.[89] The Cretans must have attacked the mercenaries at close range to be right upon their heels if they were able to 'cut them down' and enter and capture the gates before the mercenaries could close them upon their pursuers. Besides, if the Cretans shot from a distance, there would be no reason for the mercenaries to throw away their shields to escape them. On the contrary, the shield would be the only defence against the missiles. The Cretans must have been so close that the mercenaries had to drop their shields so that they might run as fast as possible to try to escape the Cretans.

Nimble Cretans

In the heat of battle, Cretans relied on their agility and speed to keep them safe, along with an exceptional endurance. In Plato's *Laws*, Clinias the Cretan specifically states that 'we Cretans are runners, since this land of ours is rugged and more suitable for the practice of foot-running', and that these were the reasons for avoiding heavy armour and adopting bows, 'because of their lightness'.[90] In the *Anabasis*, Xenophon mentions the Cretans' ability in the long-distance race during the games held by the Greeks.[91] Cretans were renowned for their speed. Livy mentions the remarkable ability of Philip V's Cretans, who were able to keep up with the horses when the king set out to cut off the Roman foragers from their camp:

> 'When he [Philip] saw them well scattered, he set out with all his cavalry and the Cretan auxiliaries, in so far as these fast-moving infantrymen could keep up with the cavalry, and marching at full speed set up his standards between the Roman camp and the foragers.'[92]

At the Battle of Magnesia, the speed and agility of the Cretans were highlighted once again when the charges of the scythed-chariots 'were easily avoided' by the 'the swift-footed Cretans'.[93] It thus comes as no surprise that Agis chose the 'Cretan runners' in his army and sent them in all haste to warn the city of Sparta against the approaching army of Epaminondas.[94]

Their speed ties in well with their ability to operate in difficult terrain. As we have seen, the uneven terrain of Crete led to the islanders being excellent light infantry. As far back as the *Anabasis*, the Cretans were noted for their fighting ability in difficult countryside. When fighting the Carduchians in the mountainous passes, for instance, Xenophon states that in 'these regions the Cretans made themselves exceedingly useful'.[95] There is also the occasion where the Cretans in Xenophon's army were used to lure the pursuing Persians away from the main Greek army. Once the Persians took the bait, the Cretans relied on their speed and agility to escape through the uneven terrain of the mountains by plunging 'out of the road into the woods, and by tumbling down through the ravines'.[96] Impressively, even while being pursued by the Persians, the Cretans were able to fight a running battle, turning to shoot arrows at their pursuers.

During a major skirmish between the Romans and the Macedonian forces of Philip V, Livy mentions that neither the Thracians nor Philip's phalanx were of any use to him due to the heavily wooded terrain. The only forces that were of any service to the king were his Cretans.[97] Incidentally, this supports the idea that the Cretans shot their arrows at close range. According to Livy, the forest where the skirmish took place was too thick for the phalanx to operate or for the Thracians to wield their *rhomphaia* polearms.[98] For the Cretans to shoot arrows at their enemies in such thick woods means they had to shoot at close range, as arrows travelling any substantial distance would find themselves obstructed by the dense foliage.

This ability also allowed the Cretans to excel in the confines of urban warfare. We have seen how the Cretan mercenaries of Demetrius II used the high ground and their range ability to take on a much larger force during the Antiochian Riots and the Cretan Tyranny. Similarly, the Cretans in the army of Cleomenes scaled the walls and rooftops of Argos, and from there sniped down at their enemies, clearing the city streets.[99]

Countering Cavalry

Archers as a means to counter cavalry have already been discussed. This was very much the case for Cretan bowmen as well. At a battle outside of the walls of Argos, Xenophon specifically records that had the Cretans of Agesilaus not been absent, 'many men and horses would have been shot down by their arrows'.[100] Plutarch mentions that Pyrrhus' horse was shot dead from under him by a Cretan at the siege of Sparta.[101] We are also told by Livy that Philip V's Cretans rendered great service to the king against the Roman cavalry, shooting 'their arrows against unprotected horses and riders'.[102] I have already mentioned that Cretan arrowheads tended to be large, heavy and barbed, not unlike the 'swallowtail' heads of the medieval era (**see plate 31**). These would be effective against exposed horses, as the heads would cause gaping, bleeding wounds. One can imagine how an arrow embedded in the flesh of a horse would grind open an even larger wound as the horse galloped across the battlefield. This is reflected in Livy's account of Magnesia, where Eumenes' Cretans and other light troops wounded the horses of Antiochus' scythed chariots and drove them back onto their own lines:

> '[T]he wounds made by the missile weapons thrown from every quarter, and partly by the discordant shouts raised, so terrified the horses, that immediately, as if unbridled, they galloped about at random.'[103]

Their fame as runners and their 'light' nimble fighting style made them ideal to serve as support units for cavalry, and we often see Cretans being paired with cavalry contingents. Since cavalry tended to fight against other cavalry, it made a lot of sense to have Cretans as anti-cavalry support.

6.4 Conclusions on the Cretans

Cretans undoubtedly played an important role as mercenaries in the Graeco-Roman world. Warfare was at the very core of Cretan society, which was based around a martial ethos embraced by an aristocratic warrior citizen class. They provided a highly specialized skill in archery, the bow being considered their national weapon. The geography of Crete made natural light infantrymen of its warrior citizens, and the island's socio-political turmoil meant that there were always Cretan bowmen willing to take on the role of soldiers of fortune. In hiring a Cretan, one thus hired a well-trained, highly specialized elite light infantry unit.

Balearic Slingers

Arguably, the most famous slingers of the ancient world were those from the Balearic Islands. Ancient sources are rife with mentions of the skill and ability of the Balearic slingers. Diodorus and Polybius both state that the very name of the Balearic Islands and its people was derived from the word '*ballein*', meaning 'to hurl', and attribute this to the skill of the islands' slingers.[104] Livy writes that at the time of the Punic Wars, the Balearic Islanders used no weapon but the sling and that no other nation could match their skill with the weapon, likening their barrage of stones to a 'violent hailstorm'.[105] Strabo writes that the Balearic slingers were 'spoken of as the best slingers',[106] while Vegetius goes as far as crediting the Balearic Islanders with the invention of the sling.[107]

Ancient authors such as Strabo, Diodorus, Florus and Vegetius all attribute the skill of the Balearics to their training with the weapon from a very young age.[108] Diodorus perhaps describes it the best:

> 'And they are so accurate in their aim that in the majority of cases they never miss the target before them. The reason for this is the continual practice which they get from childhood, in that their mothers compel them, while still young boys, to use the sling continually; for there is set up before them as a target a piece of bread fastened to a stake, and the novice is not permitted to eat until he has hit the bread, whereupon he takes it from his mother with her permission and devours it.'[109]

Their martial skill earned them a fearsome reputation and made them valuable and sought-after mercenaries. Balearic slingers served as specialist troops in both the Carthaginian and Roman armies. Carthage hired Balearic slingers by their thousands. Apparently, when Carthage first started hiring Balearic mercenaries, the islanders had no monetary system, so the Balearics in Carthaginian employ were paid in women and wine.[110] By the time of the Punic Wars, this was in all likelihood no longer the case and the soldiers were paid in coin. Indeed, we are told that Mago, Hannibal's younger brother, was sent to the islands with large amounts of coin to hire mercenaries.[111] As early as the fifth century BC, in preparation for his attack on Acragas (407–406 BC), Hannibal (not the famous Hannibal who crossed the Alps with his elephants) sent recruiting officers to the Balearic Islands to enlist slingers.[112] In 311 BC, 1,000 Balearic slingers were employed by the Carthaginian general Hamilcar against the Greek forces of Agathocles.[113] Hannibal Barca made extensive use of Balearic slingers in his preparations for the Second Punic War. He sent a force of 870 Balearics to support the garrison forces at Carthage and gave his brother Hasdrubal a contingent of 500 Balearic soldiers to reinforce Spain.[114] A large force of Balearic slingers also accompanied Hannibal on his epic journey over the Alps and his invasion of Italy. At the Battle of the Trebia (218 BC), a force of 8,000 light infantry composed of *lonchophoroi* and Balearic slingers acted as a screening force for the Carthaginian heavy infantry.[115] The Balearics were also present at the ambush of Lake Trasimene (217 BC).[116] Again, the Balearic slingers and other light troops initiated the opening stages of the devastating Battle of Cannae (216 BC), where Hannibal annihilated a Roman army twice the size of his own. In 205 BC, during the latter part of the Second Punic War, as many as 2,000 Balearics were sent to Carthage in preparation for its defence against the Roman invasion.[117] Finally, the Balearics fought in the front ranks at Hannibal's final stand at Zama (202 BC).

As Rome's dominance spread over the Mediterranean world, it brought the Balearic Islands under their control and provided access to these skilled slingers. Balearic slingers served in Roman Republican armies as auxiliaries rather than mercenaries.[118] Sulla used Balearic slingers in his war against the Numidian king Jugurtha, and Balearics also served as auxiliaries alongside the legions of Caesar during his campaigns against the Gallic tribes and continued their service well into the imperial period.[119]

Today, the ancient art of slinging is alive and well in the Balearic Islands. Slinging is Majorca's national sport, with well-defined rules and regulations.[120] Keeping with tradition, slings must be made from animal or plant material, and only stones are allowed to be used for missiles. The popularity of the sport has grown to such an extent that in 2011, Majorca held the first international sling competition, which drew competitors from nine countries.[121] There are numerous slinging clubs, and slinging is even part of the sports curriculum of many schools.

6.5 Arms of the Balearics

6.5.1 The Sling

As mentioned, the sling was very affordable and easy to construct. Due to the sling's affordability and size, it was possible for almost any man to arm himself with such a weapon. We have also seen how this was emphasized in *De Re*

Militari, which recommended that every soldier 'without exception' should be instructed in the use of the sling, for it 'cannot be reckoned any incumbrance, and often is of the greatest service'.

Balearic slingers carried up to three slings, varying in length, which were used to cast at longer and shorter distances.[122] When not in use, the slings were tied around the head or the waist.[123] According to Strabo, these slings were braided from black-tufted rush (a type of reed), hair or sinew.[124] Vergil states that the Balearic sling was braided of flax.[125] The use of braided slings of hair or flax continued into the late Roman era, as attested by Vegetius.[126] Livy, however, states that the Balearic sling was of a single piece of leather.[127] What is meant by this is that the sling was what is known as a split sling. It was a continuous piece plaited out of the same material, which simply formed a small loop or eye (hence the term split pouch or split sling) in the centre of the sling in which to place the stone, as opposed to a pocket made separate from the arms and often made of a different material (**see plate 16**). Livy continues by saying the sling was inferior in construction and accuracy to those of the Greeks from Aegium, Patrae and Dymae.[128] This appears to be a cultural bias against the Balearics, whom Livy regarded as 'uncivilized', as most ancient authors praise the skill and accuracy of the Balearic slingers and for centuries they were some of the most valued mercenaries and auxiliaries. Elsewhere, Livy contradicts his own statement, saying that 'no nation comes near them [the Balearics] in the skill with which they handle it'.[129]

6.5.2 Slingshot

A variety of different ammunition or slingshot could be used, such as simple stones, baked clay and shot cast of lead. From the ancient sources, it appears that stones remained the ammunition of choice among the Balearics throughout the Hellenistic and Imperial Roman periods, as lead or clay slingshot are never mentioned in a Balearic context.[130]

It is, however, possible that Balearics used clay slingshot as well. Numerous examples of clay slingshot have been excavated around and on the Byrsa acropolis in Carthage, but it is impossible to know whether these were used by the Balearics or other Carthaginian slinger units (**see plate 17**).[131] Being able to shape the clay would give it an aerodynamic advantage over a simple smooth stone.

Lead slingshot was the standard ammunition of the Greek and Roman slingers (**see plate 18**). Through exposure to these cultures, the Balearics might also have adopted lead slingshot to augment their slinging arsenal. As mentioned, Balearics used different slings for different sizes of slingshot. For their long-range cast, lead would have been optimal, and I find it highly probable that these elite slingers would have adopted lead slingshot.

The one element that set the Balearics apart from their slinger counterparts was surely their skill in being able to cast stones of a much larger size than other slingers.[132] According to Diodorus, these stone slingshots weighed as much as a *mina*.[133] The weight of the *mina* depended on the period and even area. The Attic/Euboic *mina* weighed 433 grams, which was based on the drachma weighing 4.33 grams.[134] The Ptolemaic *mina* weighed 490 grams, while the Roman Imperial *mina* was 337.5 grams, although the latter was only established after AD 64 when Nero reduced the weight of the *denarius* upon which the weight was based to 3.375 grams. Prior to this, Rome used the *denarius* which had a similar weight to the Attic drachma, and thus the same Attic *mina* of 433 grams. Since Diodorus wrote in the first century BC, it is safe to assume that the *mina* used by Diodorus was based on the Attic/Euboic standard of 433 grams. Due to the size of these stones, only a limited number of them could be carried.

Being able to cast a stone of almost half a kilogram with a sling is a remarkable feat, and being hit by a missile of this size would have a devastating effect (**see plate 32**). Diodorus stresses the Balearic slingers' skill in casting these immense stones:

> 'In the business of war they hurl much larger stones than do any other slingers, and with such force that the missile seems to have been shot, as it were, from a catapult; consequently, in their assaults upon walled cities, they strike the defenders on the battlements and disable them, and in pitched battles they crush both shields and helmets and every kind of protective armour.'[135]

Iconography of slingers is very rare, but there is one depiction of a slinger that portrays what is believed to be a Balearic slinger. A slinger portrayed on Trajan's Column is shown slinging a massive stone.[136] The size of the stone undoubtedly fits the description of a *mina*-sized slingshot, so it is very possible that the slinger depicted is from a Balearic auxiliary unit serving in the army of Trajan. It is interesting to note that the slinger carries his spare stone in the fold of his cloak. The ammunition, as would be expected, is fairly limited because of its size.

6.5.3 Defensive Equipment and Secondary Weapons

As with most range light troops, Balearic slingers relied on their speed and dexterity to keep them safe. They thus carried no body armour. Strabo does, however, mention that these slingers entered battle wearing no armour but for a goatskin draped around their non-slinging arm.[137] This would obviously serve as rudimentary protective equipment against enemy missiles. The slinger depicted on Trajan's Column is shown without any body armour, wearing only a short tunic, although in his left hand he carries a small round shield. If the slinger does indeed hail from the Balearic Islands, it is possible that shields were standard equipment for many slingers. At the very least, this demonstrates that slingers could and at times did indeed carry shields as part of their defensive gear.

Their social status as range light troops and the economic limitations of their island home in all likelihood also played a role in their lack of equipment. We are told by Strabo that when the Carthaginians first met the islanders, they were armed with fire-hardened spears, and on rare occasions the more well-to-do had javelins with iron heads.[138] It is very likely that javelins also formed part of the islanders' weaponry, but it was undoubtedly for their skill as slingers that they were known and for which they were sought-after mercenaries. The slinger depicted on Trajan's Column also has a dagger hanging from his belt.

6.6 Balearics at War

Multipurpose Slinger

We have already mentioned that the sling had an unequalled range ability and that Balearic slingers carried different slings in order to cast different size slingshot at different distances. The Balearic slingers were thus multipurpose slingers who could adjust to the particular need of the battlefield situation. They could respond with lighter ammunition when a longer range was required or with larger heavier stones when the circumstances required a crushing force of impact.

Anti-Armour Unit

From the literature, it is clear that the one thing that truly stands out about the Balearic slingers was their ability to cast stones of immense size. For example, Diodorus states: 'In the business of war they hurl much larger stones than do any other slingers.'[139] These massive stones would have a much shorter range but would more than make up for it in hitting power. The force was so immense that Diodorus compares it to as if the stone had 'been shot, as it were, from a catapult'.[140] The sheer weight of these *mina* stones had the ability to 'crush both shields and helmets and every kind of protective armour'.[141] The effectiveness of these slingers and their *mina* stones is best described in Diodorus' account of the Battle of Himera. During the Seventh Sicilian War (311–306 BC), Agathocles, the tyrant of Syracuse, attacked the Carthaginian camp near the mouth of the Himera River (the modern-day Salso). His Greek hoplites managed to throw down the palisade and push back the Carthaginian heavy infantry, who were getting the worst of the fight:

> 'But when Hamilcar saw that his men were being overpowered and that the Greeks in constantly increasing numbers were making their way into the camp, he brought up his slingers, who came from the Balearic Islands and numbered at least a thousand. By hurling a shower of great stones, they wounded many and even killed not a few of those who were attacking, and they shattered the defensive armour of most of them. For these men, who are accustomed to sling stones weighing a mina, contribute a great deal toward victory in battle, since from childhood they practise constantly with the sling. In this way they drove the Greeks from the camp and defeated them.'[142]

The Balearic slingers not only saved the Carthaginian camp from Agathocles' army, but completely routed the Greek forces, resulting in a major Carthaginian victory. The force with which these slingers threw their slingshot was enough to crush the heavy armour of the Greek hoplites.

Because of the immense size and weight, the *mina* stones used by the Balearics would have a much-reduced range compared to lighter stones and lead slingshot. The distance was clearly not the aim, but rather the devastating impact they delivered over a short distance. Having experimented with slinging stones weighing a *mina*, I would suggest that they had an accuracy range of 15–20 metres. The literary account of the Battle of Himera supports the idea that these large stones were cast at short distances. Since the two heavy infantry phalanxes were engaged in a melee, the Balearics had to carefully pick out individual targets before they could cast their stones. Their aim had to be precise so that they do not hit their own heavy infantry who were in close-quarter battle with the Syracusan hoplites. As with archery, the longer the range of the throw, the longer it took the missile to reach its intended target and the less likely it would be that the target would be in its original position, especially in a battle scenario.

Siege Warfare

The Balearics were also adept at siege warfare. When attacking cities, their marksmanship allowed them to clear the battlement of defenders. According to Diodorus, during 'their assaults upon walled cities, they strike the defenders on the battlements and disable them'.[143] It has already been mentioned that Diodorus compared the Balearic slingshot of a *mina* to that of a catapult. It is also interesting to note that of the four different weights of artillery stones found at the siege site of Numantia, a *mina* weight was the lightest artillery stone. For slingers to throw slingshot the same size as light artillery gives context to Diodorus' comparison of these slingers with catapults. It was perhaps their ability to throw such large slingshot that contributed to their effectiveness in siege warfare. The Carthaginians used their slingers, who included large numbers of Balearics, to great effect in their siege of Selinus (409 BC) 'to drive back the combatants on the battlements'.[144]

The Balearic slingers' skills were also effective in defending during siege attacks. During Caesar's Gallic campaigns, Bibrix, an allied city, was besieged by the Belgae. The city was completely surrounded by a large force of warriors who had a great number of slingers. The Belgae slingers were highly effective at clearing the city's battlements of defenders:

> '[S]tones have begun to be cast against the wall on all sides, and the wall has been stripped of its defenders, [then], forming a testudo, they advance to the gates and undermine the wall: which was easily effected on this occasion; for while so large a number were casting stones and darts, no one was able to maintain his position upon the wall.'[145]

The city was only saved by nightfall, when the assault was brought to a halt. A messenger was sent from the city to request immediate aid from Caesar, who responded by sending several units of Numidians, Cretan archers and a contingent of Balearic slingers. These reinforcements were able to defend the city effectively enough for the Belgae to abandon the siege.

6.7 Conclusion on Balearic Slingers

The sling was more than the national weapon of the Balearics: it was a way of life. From childhood, the islanders were taught the art of slinging, and if the literature is to be believed, they did not eat if they did not master the skill. In a mainly pastoral society, the sling was a tool for everyday life, being used for both the guidance and protection of their flocks. Slinging was thus very much a part of the islanders' life. This skill made them sought-after mercenaries, particularly in the Western Mediterranean world.

They were not only masters of the sling but were also skilled at using different slings for a variety of stones and for different distances, giving them a level of diversity that was lacking in most other slinging units. A particular element that set them apart from other slingers was their ability to cast large stones weighing up to a *mina* (433g) in weight. This is particularly highlighted in Chapter 8's experimental archaeology section (see Chapter 8: Slings & Slingshot). It took tremendous skill and practice to cast stones of such size, and there was a reason why the Balearics 'in the business of war … hurl much larger stones than do any other slingers'. This ability made them an effective anti-armour unit. Even the heaviest armour could not withstand the impact of such a large stone flung from a Balearic slinger, as events at the Battle of Himera clearly showed.

They also filled much-needed range combat roles in the Western Mediterranean. While javelinists featured prominently in Western armies, as discussed, archery was not as extensively practised in the Western Mediterranean as it was in the East. This severely limited the range ability of Western armies. Having such an effective slinger unit was therefore vital to augment the range capabilities of Carthaginian and Roman armies.

Chapter 7

Battle Assessment

Epic battles such as Marathon, Thermopylae, Cannae and Zama, where the heavy infantry carried the day, are very well known. Yet history is full of lesser-known or even well-known battles where the light infantry dominated the battlefield. In these encounters, the light infantry played the most prominent roles and were the true battle winners. Many of these battles were so decisive that they not only defeated the enemy but had a profound influence on the war. In some cases, these battles won the entire war, further emphasizing the effectiveness of light troops on the battlefield.

Many of these victories were achieved because of the topography of the battlefield, the lack of adequate light infantry support on the part of the defeated, a general's ability at using combined-arms tactics or his ability to capitalize on the skills of light infantry when the enemy failed to do so.

This chapter will assess some of the most prominent battles of the Graeco-Roman world where the light infantry played a crucial role on the battlefield and their actions were essential for victory.

7.1 Sparta's Disastrous Defeat at Sphacteria, 425 BC

7.1.1 The Battle of Sphacteria

The Battle of Sphacteria was one of the most important battles of the early phase of the Peloponnesian War, known as the Archidamian War (431–421 BC). The battle was the result of Athenian general Demosthenes capturing Pylos in response to Sparta's repeated invasions of Attica. Situated on the west coast of the Peloponnese, Pylos was to serve as an Athenian base of operations to raid Spartan territory. A failed Spartan attempt to recapture Pylos resulted in a contingent of Spartan hoplites being trapped on the island of Sphacteria by the Athenian fleet.

The command of the expedition against the Spartans on the island was given to Cleon after he boldly claimed to the Athenian assembly that he had no fear of the Spartans and that he would set sail without taking any hoplites from the city. He thus took with him only contingents from Lemnos and Imbros, a number of *peltasts* and archers. Upon arrival, he added to his forces the troops stationed at Pylos. Cleon chose as his colleague for the expedition the experienced Demosthenes, who was the real strategic talent behind the great Athenian victory. Demosthenes had already planned an attack on the island prior to Cleon being chosen as commander of the expedition, as the troops under him were getting weary of the limited supplies and the inactivity after their victory at Pylos.

The island, having never been inhabited, was covered in woodland, and the position of the Spartan contingent was well hidden. The Athenians were wary of landing their forces because of the thick vegetation and the possibility of a surprise attack due to the difficulty of tracking the movements of the Spartans:

> 'The fact that the island, having never been inhabited, was largely wooded and pathless had up till now deterred him [Demosthenes], as he thought this gave an advantage to the enemy. Even if he landed with a large force the Spartans could do damage with attacks from hidden positions. The tree-cover would deny the Athenians a clear view of the enemy's shortcomings and capabilities, whereas any false move by their own forces would be plain to see, so the enemy could make unpredictable attacks wherever they wished, and the initiative would be theirs. Then again, if he were compelled to close quarters in a wooded area, he thought that the smaller force with a knowledge of the ground would have the advantage over the larger force without that knowledge: and with no sightlines to indicate where support was needed the Athenian army, large though it was, could be destroyed without realizing the extent of the destruction.'[1]

Luckily for the Athenians, a wildfire broke out on the island, burning away much of the vegetation and woodland, revealing the best landing spots and exposing the Spartan position. Not only did the fire show the Spartans' location, it also revealed the number of soldiers – 440 hoplites and their helot servants, for which Thucydides gives no number. The Spartan force on the island occupied three different positions. The largest group under the Spartan commander Epitdas was stationed in the centre of the island around a spring, which served as their water source. The second position lay in the north and was occupied by a smaller contingent. This was the island's highest point, today known as Mount Elias, the remains of an old fort crowning the high ground which overlooked the Athenian camp. The elevation, the protection provided by the remaining walls of the fort and the steep cliffs on its eastern side made it a highly defensible position. It was deemed by the Spartans to be a good place to make a stand if they were hard-pressed. The third division, which was the smallest of the three with approximately thirty hoplites and their helot attendants, was stationed to the south of the island to watch against possible landings there made by the Athenians.

The Athenians had originally underestimated the Spartan force, and even though Cleon and Demosthenes had a much larger contingent of both heavy and light infantry, the fierce Spartan martial reputation made them very cautious. A herald was sent to the Spartans on the mainland to instruct the soldiers on the island to surrender, which, in true Spartan fashion, was promptly rejected. The next night, under the cover of darkness, the Athenians landed their vanguard on the western and eastern sides of the island. They quickly overran the small Spartan force stationed to the south, which was wholly unaware of the enemy's arrival, with many perishing while still in their beds:

> 'The Athenians rushed the first guards and killed them instantly – they were still in bed or grabbing for their weapons. The landing had gone unnoticed, as the guards had thought that the ships were simply moving to their usual nightly anchorage.'[2]

At dawn, the remainder of the Athenian forces disembarked on the shores of Sphacteria. The Athenian force consisted of 800 hoplites, 800 *peltasts* and 800 archers. In addition to these were the crews of the seventy triremes, who would serve as *psiloi*, with the exception of the '*thalamioi*', the lowest-tier oarsmen, who stayed with the ships. Demosthenes' strategy was to avoid a heavy infantry engagement at all costs, even though the Athenians had a superior number of hoplites, thereby playing against the Spartans' strength of hoplite warfare. He divided his army into contingents of more or less 200 men, and with these occupied the high ground around the main Spartan camp. The Spartans were now completely surrounded:

> '[The Spartans] being completely surrounded and distracted by the number of their opponents, might not know whom they should face first, but might be exposed to missiles on every side. For if they attacked those who were in front, they would be assailed by those behind; and if those on one flank, by those posted on the other; and whichever way they moved, the light-armed troops of the enemy were sure to be in their rear.'[3]

Even now, where the effectiveness of the light troops was at its pinnacle, Thucydides pays these second-class soldiers a backhanded compliment by highlighting the embarrassment of being defeated by them:

> 'These were their most embarrassing opponents, because they were armed with bows and javelins and slings and stones, which could be used with effect at a distance. Even to approach them was impossible, for they conquered in their very flight, and when an enemy retreated, pressed close at his heels. Such was the plan of the descent which Demosthenes had in his mind, and which he now carried into execution.'[4]

The effectiveness of Demosthenes' strategy and the dire situation the Spartans found themselves in is best described by Thucydides:

> 'When the contingent under Epitadas [the main body of the Spartans on the island] saw that their first garrison had been wiped out and an army was moving against them, they formed for battle and advanced towards the Athenian hoplites, wanting to come to close quarters: the hoplites were positioned directly in front of them, with the light-armed troops on the flanks and behind them. The Spartans could not get at the hoplites or make use of their own expertise, as they were kept back by a barrage of missiles from the light-armed troops on either side, and also the hoplites would not advance to meet them but stayed where they were.'[5]

In contrast to the conventional military strategy of the time, it was the heavy infantry who played the supporting role and the light troops who took the most active and decisive part in the battle.

The heavy armour of the Spartans, which was meant to protect them in battle, now became their bane. Whenever they were attacked from a particular side, the Spartans responded by sallying out from their formation and driving off the light troops. Their heavy shields and gear, however, made it impossible for them to catch their lightly armoured and unencumbered foes. The fact that the battle took place in rough terrain only made the situation worse for the Spartan hoplites:

> 'They could drive back the light-armed wherever they ran in particularly close to make their attacks, but then these troops would turn again and renew the fight: being lightly equipped they could easily outpace any pursuit in this difficult terrain, left rough by lack of habitation, and the Spartans in their heavy armour could not chase them over such ground.'[6]

Fatigue from the continuous fruitless sallies and the numerous wounds suffered started to wear down the Spartans. More importantly, the image of the invincible Spartan hoplite was being shattered. The Athenian light infantry was being emboldened by every arrow, javelin and stone that hit its mark. Casualties mounted among the Spartans, while the light troops launched repetitive, ever more daring attacks and retreated unscathed:

> 'For some little time they continued this long-range skirmishing against each other. But when the Spartans were no longer able to respond with quick sallies wherever they were attacked, the light-armed troops could see that they were now slower to defend themselves, and this sight had greatly increased their own confidence. They could see, too, that they clearly outnumbered the Spartans, and with greater experience of them now did not think them as formidable as they once had: they had made the initial landing in a state of abject terror at the thought of facing Spartans, but had suffered nothing so far to justify their apprehension. Turned now to contempt, they gave a shout and launched themselves in a body against the Spartans, bombarding them with stones, arrows, javelins – whatever each of them had to hand.'[7]

In addition to the difficult terrain, dust and ash from the wildfire rose up, disturbed by all the movement of the opposing armies. This made it extremely difficult for the Spartans to see the 'showers of arrows and stones hurled by their assailants which were flying amid the dust'.[8] The situation was exacerbated by all the shouts and screams of the enemy which 'dismayed the Lacedaemonians, who were unaccustomed to this kind of warfare … not being able to see out of their eyes or to hear the word of command, which was drowned by the cries of the enemy'.[9]

Being attacked from every direction, 'the Lacedaemonians began to be sorely distressed'.[10] They realized their *pilos* helmets could not 'protect them from the arrows', and the 'javelins broke off where they struck them'. I have mentioned that some scholars interpreted the *pilos* to be made of felt (the word '*pilos*' refers to the method of compressing wool to make felt), while others have believed it to mean felt cuirasses. It more likely refers to the metal helm called a *pilos* which was shaped in the same way as the felt cap and took its name from the helmet. Being a completely open-face helm, covering only the head meant that it was far easier for arrows to hit their mark. Experimental archaeology has demonstrated that even if the Spartans wore bronze cuirasses, a javelin thrown with the *ankylē* throwing strap produced enough kinetic energy to penetrate the armour and pierce the wearer's flesh. Thucydides' comment that the 'javelins broke off where they struck them' can thus be understood as the tips penetrating their bodies, yet the wooden shafts snapped under the force of impact.

Finally, 'destruction was staring them in the face, and they had no means or hope of deliverance', so the exhausted Spartans closed their ranks in a tight formation for defence, possibly a square formation of hoplites with the helots in the centre, similar to that used by Brasidas in 423 BC against a large force of Illyrian *psiloi*. They fell back to the remnants of the fort on the high ground to make their last stand. Hard pressed by the light troops, the Spartans lost a few more men during their tactical retreat, but when they finally reached the fort, which was not too far off from their main position, they received some respite from the incessant attacks. The restricted space, the walls and the natural defence provided by the high ground and the cliffs made it possible for the Spartans to effectively defend an attack from all sides. The Spartans were now also reinforced with some fresh troops, those who were originally stationed at the fort, and in all probability the Athenian missiles were running low. The few helots among the Spartans who acted as *psiloi* also had some advantage in casting their missiles from the high ground. The most important factor in the defence was the fact that the Spartans could no longer be outflanked and attacked in the rear or their vulnerable flanks.

The battle had by now developed into a stalemate. The Athenians could not dislodge the Spartans from their fortified position, but the Spartans had no means of driving off their assailants:

> 'For a long time, indeed for most of the day, both sides held their ground, enduring the fighting and thirst and the heat of the sun, one side determined to drive the enemy from the high ground, the other determined not to give way. But it was now easier for the Spartans to put up a defence than before because their flanks were no longer unprotected.'[11]

The commander of the Messenian troops within the Athenian force, realizing the futility of the current situation, requested Demosthenes and Cleon to give him some archers and other *psiloi*. With these, he would circumnavigate the Spartan position, scale the difficult cliffs in their rear and attack the Spartans from behind. Thucydides tells us that it was only with great difficulty that the commander of the Messenians managed to scale the precipitous cliffs. He eventually succeeded, and when he appeared with his forces on the high ground behind the Spartans, it caused great terror and panic among the Lacedaemonians, meanwhile instilling the Athenians with fresh vigour and lifting their spirits.

Thucydides now compares the battle to that of the great Battle of Thermopylae (480 BC), where the Spartans had also been outflanked and attacked in the rear by the Persians:

'The Spartans were now under attack from front and rear and were in the same dire situation, to compare small things with great, as the men at Thermopylae. Just as those men were annihilated when the Persians outflanked them by taking that path, so these Spartans could not hold out any longer, caught as they were between two fires. They were few fighting many and weakened by lack of food, and they began to give way with the Athenians controlling every approach to their position.'[12]

Both Cleon and Demosthenes realized that if they did not call off their troops, the entire Spartan force would be annihilated; and the Spartan soldiers were far more valuable alive than dead. A herald was then sent to the Spartans to offer them a chance to lay down their arms and surrender. In an unprecedented turn of events, the Spartans accepted the offer of surrendering and a truce was agreed upon.

Epitadas, the commander of the Spartan forces, by now lay dead on the battlefield, and his second-in-command, Hippagretus, was badly wounded, serving as a grim example of the devastation inflicted upon the Spartan heavy infantry and the effectiveness of the opposing light troops. Styphon was thus left in command, and he negotiated the armistice with Cleon and Demosthenes. During the negotiations, Styphon requested to consult with the Spartan forces on the mainland on how to proceed. The Athenians refused to let any of the Spartans leave the island, but allowed emissaries to cross over from the mainland. These told Styphon that 'the Spartans tell you to make your own decisions about yourselves, but to do nothing dishonourable'.[13] Exhausted, low on supplies, completely surrounded and severely outnumbered, the trapped Spartans had to choose between certain death and complete surrender. They chose the latter.

7.1.2 Battle Assessment

The Battle of Sphacteria had a great impact on the war, sending shockwaves through the Greek world. The Spartans had stationed a force of 420 hoplites on the island; 292 of these survived, the rest all perishing under a hailstorm of stones, javelins and arrows. Of the survivors, 120 were Spartiates of the elite warrior class. Thucydides fails to mention the number of helot servants who attended their Spartan masters, nor does he mention the number of their dead. Nevertheless, 'not many Athenians were killed because the battle was not fought at close quarters'.[14]

Not only had Demosthenes and Cleon beaten the 'unbeatable' Spartans, but they had done so with light infantry, who were considered second-class soldiers. More significantly, Sparta's military reputation was severely tarnished:

> 'To the Greeks this was the most surprising event of the whole war. They had thought that Spartans would never surrender their arms, in starvation or any other extremity, but would use them to the last of their strength and die fighting.'[15]

When Sparta entered the war against Athens, it had expected to quickly overcome its adversary and further cement the Spartan hegemony over Greece. The defeat on the island of Sphacteria awoke Sparta to a brutal reality:

> 'The war was not going according to plan, as they had expected to break Athenian power within a few years if they ravaged their land; they had met with the disaster on the island, something unprecedented in Spartan history.'[16]

From here on, the war took a different turn, with power shifting to the Athenians, who were emboldened by their victory. No longer would Attica suffer the devastation of Spartan invasions and raids; if so, the prisoners taken from Sphacteria would be executed. For the first time since the start of the war, Attica's crops were safe. In contrast, it was now Laconia that was exposed to raids from the newly created base of operations at Pylos, which was manned by the Messenians, who 'were able to cause a great deal of damage' to the Spartan lands.[17] Furthermore, the defeat threatened to destabilize Spartan society: 'finding the Helots deserting, and dreading some serious domestic calamity, they were in great trouble', the Spartans fearing 'the possibility of a revolution overtaking their political system'.[18]

The Spartans found themselves on the back foot, the defeat at Sphacteria dealing a severe blow to their morale and confidence:

> '[T]hey now feared some further disaster like that which they had sustained on the island. This made them less confident for battle. Unused to reverses before now, they had lost faith in their judgement and thought that any move they made would end in failure.'[19]

Sparta's leadership also realized that they could simply no longer rely solely on their heavy infantry, as brave and highly trained as they were, to fight the war. Consequently, the Spartans expanded and diversified their army: 'Contrary to their usual custom they raised a force of four hundred cavalry and archers.'[20] The battle had brought an end to the hoplite domination; while they were still considered the most valuable soldiers on the battlefield, the events on the island highlighted the severe limitations of the hoplite heavy infantry. Sphacteria proved that the best-trained, most disciplined army of heavy infantry could easily be overcome by the very ones whom they deemed inferior to themselves, both in status and in fighting style. While the light troops would never be the equals of the elite hoplites, their relevance on the battlefield could no longer be ignored and their deployment became of the utmost importance.

7.2 Antiochus' Crossing of Mount Labus, 210 BC

7.2.1 The Battle of Mount Labus

Antiochus the Great's *anabasis* saw the Seleucid king campaign in the East to re-establish dominion over the breakaway kingdoms of Parthia and Bactria. In 210 BC, the king led his army across the Elburz mountain range to invade Hyrcania, which was under Parthian control. However, Mount Labus Pass, which led into Hyrcania, was occupied by a large number of enemy forces who had fortified various sections of the pass in order to prevent the Seleucid crossing.

The flower of the Seleucid army was the heavy infantry of the Macedonian-style phalanx. In large open field battles, the phalanx was the main strength of the king's forces, Livy describing them as the 'strongest' of all the troops of the Seleucids according to Appian.[21] The phalangites were armed with the monstrous *sarissa* pike measuring 14–16 cubits (6.72–7.68 metres) in length, which required both hands to wield.[22] This meant the smaller round *peltē* shield, some '8 palms' (64cm) in diameter, was attached to the left forearm for defence.[23] Because the fighting style of the phalanx relied on a highly structured formation of serried ranks, its members were well-trained and disciplined fighters. The reliance on its well-maintained formation of ranks and files meant that the phalanx required an even, fairly flat plain of open ground to operate successfully. Polybius highlights that the phalanx was serverely restricted by battlefield terrain:

> 'It is acknowledged that the phalanx requires level and clear ground with no obstacles such as ditches, clefts, clumps of trees, ridges and water courses, all of which are sufficient to impede and break up such a formation. Everyone would also acknowledge that it is almost impossible except in very rare cases to find spaces of say twenty stades or even more in length with no such obstacles.'[24]

The march through Mount Labus Pass was anything but 'level and clear ground with no obstacles'. Besides the usual difficult terrain of the mountain pass and 'quantities of rock and trees that had fallen of their own accord from the precipices above', the advance of the Seleucids was made even more treacherous and difficult by the obstacles placed throughout the pass by the enemy:

> '[N]umerous other obstacles placed there by the barbarians contributed to the result. For they had constructed a series of barricades of felled trees and had collected a quantity of huge rocks, while they themselves along the whole defile had occupied favourable positions on the heights where they fancied themselves in security.'[25]

While the Macedonian-style phalanx had its 'characteristic virtue and strength [and] nothing can sustain its frontal attack or withstand the charge', in the uneven and difficult terrain of the pass, the formation became less than useless as a fighting force. The phalanx was extremely vulnerable while in a marching column. The baggage train, the very lifeline of a campaigning army, was just as vulnerable in the hostile pass. Polybius comments that 'Antiochus would have found it perfectly impossible to execute his project' of crossing the mountains had it not been for his light troops.[26]

Fortunately for the king, he had an experienced and versatile army that consisted of well-trained and diverse unit types. An experienced general at this point of his career, and having a clear understanding of the dangers posed by the pass, Antiochus divided his '*euzōnoi*' (Polybius' preferred term for light infantry) into several divisions under the overall command of his general, Diogenes. These divisions of light troops, consisting predominantly of archers (*toxotai*), slingers (*sphendónē*) and javelinmen (*akontistai*), were to scout the various routes ahead and dislodge the enemy forces from their fortified positions. Many of his light troops were from mountainous tribes and experts of this type of irregular warfare in difficult terrain, including

> 'archers and slingers and those of the mountaineers who were expert in throwing javelins and stones, who also, whenever time and place called for it, fought singly and rendered most useful service on difficult ground.'[27]

The sophistication of the Seleucid army is demonstrated by the fact that it was accompanied by engineers and pioneers whose duty was to clear a path so that the baggage train and heavy troops of the phalanx could move through the pass.

At the head of the column, Antiochus placed his 2,000 Cretans as a vanguard, providing vital protection to the phalanx and baggage train, as well as the engineers. Their light equipment and experience fighting in difficult mountainous terrain made the Cretans ideally suited for the task. Furthermore, the Cretans provided support to the various light troops moving along the heights by 'covering their movements and marching parallel to them close to the defile slowly and in good order'.[28] Their shields provided them with some protection from enemy missiles, while their bows allowed them to respond to the enemy's range troops.

The vulnerable rear of the column comprised Antiochus's *thōrakitai* and *thureophoroi*.[29] Their role as multi-purpose light troops was to engage in both range and melee attacks, their body armour and *thureos* shields giving them the means to counter any enemy troops who attacked the column directly.

What followed was a slow, gruelling ascent through the mountain pass, with success depending entirely upon the combined efforts of the light troops. The contingent under Diogenes scouted for routes to outflank the enemy positions. Their movements along the heights were covered by the Cretans, who were marching parallel to them in the pass. Once the light troops occupied the high ground above the entrenched enemy, they rained down a storm of arrows, javelins and stones upon them:

> 'For at once upon encountering the enemy he acted as circumstances suggested and making a further flank movement uphill got on higher ground, and by throwing showers of javelins and stones from the hand inflicted severe punishment on them, the greatest damage being done by the stones slung from a distance.'[30]

At the same time, the Cretans bombarded the enemy positions with their arrows from below. Once an enemy position was cleared, the engineers could move forward and clear the path of obstacles so that the phalanx and baggage train could advance under the protection of the *thureophoroi* and *thōrakitai*.

The process was slow and methodical, but the efforts of the light troops cleared the pass of enemy forces and allowed the baggage train and the phalanx to move safely through the pass:

> '[B]y this means, with the slingers, archers and javelineers marching along the high ground in loose order, but closing up and occupying favourable positions, and with the Cretans covering their movements and marching parallel to them close to the defile slowly and in good order, the barbarians no longer stood their ground, but abandoning their positions collected on the actual summit of the pass.'[31]

The 'barbarians', as Polybius describes them, had now abandoned the pass and collected in large numbers at the summit for a final stand against the Seleucids; they were 'convinced that they would prevent the enemy from crossing, and a fierce struggle now took place'.[32] The summit opened up into fairly open ground, which allowed for Antiochus to deploy his heavy infantry of the phalanx. However, the phalanx, fighting uphill, was unable to drive back the enemy, who in a 'dense mass fought desperately against the phalanx face to face'.[33]

To break the stalemate, Antiochus, under the cover of darkness, sent his light troops to outflank the enemy's position and occupy the high ground behind them. When the enemy realized that their rear was threatened, they 'were panic-stricken and took to flight'.[34]

7.2.2 Battle Assessment

The crossing of Mount Labus serves as a prime example of the importance of light infantry. The battle emphasized the limitations of the phalanx and how restricted its deployment was. While the phalanx was, as Plutarch describes it, 'an invincible beast as long as it is one body and can keep its shields locked [*synaspismos*] together in a single formation', it was extremely vulnerable when not in formation, which was the case when marching through the pass.[35] Had it not been for Antiochus' light troops, the king would have had to abandon the campaign and return home, forsaking all hopes of reconquering his eastern provinces. The king might even have lost his army, perhaps his own life, in an attempted crossing. The successful passage through the pass depended entirely upon the actions of the light troops, who did not suffer the same limitations in difficult terrain as the heavy infantry:

> '[T]he phalanx and pack-train could not march by any other route than the one they supposed, since it was impossible for that part of the army to attack the mountain slopes, yet it was by no means beyond the power of unburdened and light-armed troops to ascend over the bare rocks.'[36]

The combined actions of the light infantry allowed Antiochus to safely move his entire phalanx and baggage train across the treacherous mountain range into Hyrcania. The king was thereby able to successfully campaign against the Parthians and Bactrians, re-establishing Seleucid suzerainty over the breakaway kingdoms.

It is interesting to note that even when Antiochus was able to deploy his phalanx at the summit, the heavy infantry were unable to break through the enemy troops. It was only when Antiochus used his light infantry to find routes through the mountainous terrain to outflank the enemy position and attack them in the rear that the enemy broke and abandoned their defence of the mountain pass. This was in keeping with contemporary Hellenistic 'hammer and anvil' tactics, using the heavy infantry phalanx as the 'anvil' to hold the enemy in place whie outflanking them and hitting them in the rear with the 'hammer', a role traditionally filled by the cavalry. The Seleucids were famous for their cavalry, which was superior both in number and quality to the other Hellenistic kingdoms.[37] As with the phalanx, however, the cavalry could not be deployed in the difficult mountainous terrain, with the role of the 'hammer' being taken by the 'unburdened and light-armed troops', who did not suffer these restrictions.

Military campaigns of the Graeco-Roman period consisted of numerous marches across mountain ranges and through difficult terrain very similar to Mount Labus Pass. In all of these cases, it was the light troops who played the vital role of protecting the baggage train and the marching column of heavy infantry by scouting and clearing the terrain of enemy forces. Although pitched battles were still dominated by the heavy infantry, the light infantry had the essential task of getting their heavily armed and armoured compatriots safely to the battlefield. Mount Labus thus serves as a perfect example of the importance of light troops to campaigning armies of the Graeco-Roman period.

7.3 The Galatian Campaign, 189 BC – Roman Victory in the Mountains

7.3.1 The Battle of Mount Olympus in Anatolia

In 278 BC, an immense Celtic army, along with their entire families, crossed the Hellespont to serve as mercenaries for the Bithynian king, Nicomedes I. Not content to be merely hired soldiers, the Celts quickly carved out a new homeland for themselves centred on Ancyra, creating the Kingdom of Galatia. From here, the Galatians became the bane of the peoples of Asia Minor, raiding and ravaging their countryside and cities. They were eventually defeated by Antiochus I at the Battle of the Elephants (somewhere in the late 270s or early 260s BC).[38] Their power was, however, by no means broken and they would still occasionally set out on a terror spree, raiding the lands of the neighbouring peoples. These raids were so widespread and devastating for the Greek cities in Asia that a special kind of tax, *ta Galatika*, was instituted to either pay for the ransom of citizens taken in these raids or finance the defensive war against Gallic incursions.[39]

In 189 BC, Gnaeus Manlius Vulso was elected as consul for the year and took command of the armies in Asia Minor which had just defeated Antiochus the Great at the Battle of Magnesia. Eager to make a name for himself and gain as many riches as he could from his term in office, he led the armies in a campaign against the Galatians. It was the first war started by a consul without the permission of the Senate. Manlius' excuse for starting the war was that the Galatians had supported Antiochus the Great in his war against Rome. The consul also capitalized on the continuous raids made by the Galatian tribes, which he deemed a danger that needed to be brought to an end. The legions were enthusiastic for the campaign, knowing that the Galatian tribes had grown rich from the raids and there would be plenty of booty and slaves for them to seize if they were victorious.

Manlius turned to Rome's faithful ally, Pergamum, for aid in his Galatian campaign. Eumenes II, the king of Pergamum, was in Rome at the time, but his brother, Attalus, who acted as regent in the absence of Eumenes, eagerly joined Manlius in his campaign against the Galatians. Attalus provided a contingent of 1,000 infantry, mostly light troops, and 500 cavalry.[40]

The Gallic tribes decided not to fight the Roman legions in the open, instead moving their armies into the mountains to make a stand against the Romans:

> 'They [the Galatians] had adopted this plan particularly for conducting the war – that, when they had occupied the highest peaks in the region, conveying everything there which would be sufficient for their use over however long a period, they would wear down the enemy by exhaustion; for, they were convinced, the Romans would neither venture to climb over such steep and difficult ground, and, if they did attempt it, they could be stopped even by a small force or pushed back, nor would they sit quietly at the foot of cold mountains and endure chill and hunger.'[41]

The Tolostobogii and Trocmi, two of the three main tribes of the Galatians, took their position on Mount Olympus. In addition to the protection afforded by the mountainous terrain, the Galatians fortified their position and reinforced it with a ditch.

When Manlius arrived with his army, he was quick to realize that his heavy infantry legionaries would be at a major disadvantage fighting an uphill battle in difficult terrain. It was clear that he would have to rely on his light infantry, especially his range units:

> 'The consul, because he had foreseen, as a result of reflection, that the fighting would not be done hand to hand but by making attacks from a distance, had prepared a vast quantity of javelins, *velite* javelins, arrows, bullets and stones of moderate size which could be thrown from slings, and equipped with this supply of missiles he led the army towards the Olympus mountain and encamped about five miles away.'[42]

Having supplied his light infantry with ample ammunition, he sent out light cavalry scouts to survey the area. These were attacked and driven off by a Galatian cavalry force twice their number, suffering several fatalities and a large number of wounded. The next day, Manlius marched out with his entire army to reconnoitre the area to find the best possible route for an attack. None of the Galatians ventured out of their fortified position to attack such a large force, and Manlius was able to establish that there were three possible routes by which he could attack the Galatian camp.

The following day, Manlius divided his army into three columns to attack the Galatian camp along the three approaches. The light troops would be the strike force, while the heavy infantry were kept in the rear as support. The elephants and cavalry were left at the foot of the mountain as the ground proved impossible for them to navigate.

When the Galatians saw the Romans approach, being confident in their position of holding the high ground, they sent out a force of about 4,000 men to a hill overlooking the Roman approach to hinder their advance. Manlius responded by sending forward his light infantry, consisting of Roman *velites*, the Cretan archers and slingers provided by Attalus and a contingent of Thracian and Trallian auxiliaries. The battle commenced with a hail of missiles from the Romans. The Galatians responded by throwing stones and a few javelins of their own at the Romans below. The battle was at first equal, with the Galatians having an advantage because of their position on the high ground, while the Romans had superior numbers and the advantage of many different missiles. However, as the battle continued to rage, the fighting started to favour the Romans. The Galatians were not prepared for a prolonged missile battle, having expected to fight the Roman heavy infantry in a melee after their disciplined formations were broken up by the difficult terrain. Once their formations were compromised, their heavy equipment would have become a hindrance in the one-on-one combat. In contrast, the Galatians' lack of armour and light shields would have given them an advantage over their burdened foes, and they would have been able to fight more freely on the rough ground. However, their lack of armour and their narrow, oval-shaped shields, which did not curve around their bodies, proved ineffective against the large number of missiles fired at them by the Romans from a variety of angles:

> 'Their shields, long, but not wide enough for the size of their bodies and, moreover, flat, offered poor protection to the Gauls. They had at this time no other weapons than their swords, for which there was no use, since the enemy did not meet them in hand-to-hand conflict.'[43]

Since the Galatians had expected a melee battle, they did not have an adequate supply of missiles to counter the Roman light troops. With their long slashing sword useless in range fighting, they were forced to resort to throwing stones by hand. But the stones proved ineffective as they were 'not of suitable size', as Livy describes them. Furthermore, the Galatians were unskilled at throwing stones:

> 'Stones – but not of suitable size, since they had made no preparations in advance, but took each what happened to come to his hand in his hasty search – they did use, but like men untrained in their employment, with neither skill nor strength to add effectiveness to the blow.'[44]

The Galatians simply had no means to counter the cloud of missiles, with stones, slingshot, arrows and javelins directed against them. Livy gives a good description of the desperate situation in which the Galatian warriors found themselves:

> 'Arrows, sling-bullets, darts, coming from all sides wounded them unexpectedly, nor did they see what to do, as their minds were blinded by rage and fear, and they were involved in a kind of battle for which they were very ill-adapted. For, as in hand-to-hand fighting, where they can receive and inflict wounds in turn, passion inflames their minds, so when they are struck by light weapons, coming from unseen and distant sources, and when they have no place at which they can charge with blind violence, like wounded animals they rush headlong upon their own friends. The fact that they fight naked makes their wounds conspicuous and their bodies are fleshy and white, as is natural, since they are never uncovered except in battle; so that both more blood flowed from their abundant flesh and the wounds stood out to view more fearfully and the whiteness of their skins was more stained by the black blood. But they are not much disturbed by open wounds; indeed, sometimes they cut away the skin, when the gash is broad rather than deep, and think that thus they gain greater glory in the fight; the same men, when the sting of an arrow or of a bullet that has buried itself in the flesh torments them, having caused a wound small to look at, and, as they search for a way to extract the missile, it does not come out, turning to madness and shame at being destroyed by so small a thing, throw their bodies upon the ground. So in this instance they lay prostrate here and there.'[45]

In desperation, some Galatians rushed out to close with their tormentors in a melee, but this only exposed them to more missiles. The few who did reach the light Roman troops, we are told, were easily dispatched by the *velites,* who, with their *parma* shields and 'Spanish sabres', were well equipped to such fighting – 'if he is compelled to fight hand to hand, he shifts his javelins to his left hand and draws his sword'.[46] The few remaining Galatians, seeing the terrors inflicted upon their fellow warriors, abandoned the hill and retreated to their fortified camp.

When the heavy infantry, having made slow progress over the difficult ground, finally arrived, the battle for the hill was over, having been won by the light troops. The heavy infantrymen were given a moment to rest from their arduous march over the mountainous terrain, while the light troops scoured the battle site to collect ammunition for the attack on the Galatian camp. When this was completed, the Romans marched on the encampment.

The light infantry wre once again sent ahead, and when they arrived the Galatians marched out of the camp to face them. It was very much a repeat of the battle for the hill, except that the large numbers of warriors grouped together made it even easier for the Roman missiles to hit their targets. The hail of missiles drove the Galatians back into their fortified camp, where they barred the gates and a strong contingent of warriors formed up in a tight shield wall:

> 'They [the Galatians] were overwhelmed by missiles of every kind, and the more numerous and the more crowded together they were, the less did any weapon fall without effect so that in an instant they were driven back within the rampart, leaving only strong guards at the actual gates of the fortifications.'[47]

Even behind their fortifications, the Galatians were not safe. Missiles poured over the ramparts into the encampment, and 'the shouts mingled with the wails of women and children showed that many were wounded'.[48]

When the heavy infantry arrived, the front ranks cast their *pilai* javelins at the shield wall protecting the entrance to the camp. None of the soldiers in the shield wall were injured by this, but the shields of the Galatians were so tightly formed that they overlapped, and the *pilai* penetrated through both shields and riveted them together, forcing the warriors to abandon them. After this, the Galatians stopped resisting and attempted to flee. Livy tells us that many, in their dire rush to escape, fell to their deaths down the cliffs and precipices:

> '[T]he Gauls began to flee from the camp in all directions. They rushed blindly along the roads and where there were no roads; no steep rocks, no cliffs, held them back; they feared nothing but the enemy; accordingly many were killed, falling headlong or slipping in weariness over tremendous precipices.'[49]

We are told by Livy that the victorious soldiers were restrained from pillaging the Galatian camp and ordered by Manlius to pursue the Galatians, the consul 'being convinced that the war was over if the largest possible number was killed or captured in that rout'.[50] Burdened by their gear, the heavy infantry would have struggled in this task, so the light infantry would have undertaken it. Only when the Galatians reached the plain were the cavalry able to join the pursuit. What ensued was a massacre:

> 'The number of casualties could not easily be calculated because the flight and slaughter went on far and wide over all the outlying parts of the mountains, and a great number fell from the pathless cliffs into valleys of enormous depth, and some were killed in the forests and thickets.'[51]

The sources differ on the number of casualties, ranging from 10,000 to 40,000.[52] It is more likely that the prisoners and casualties together amounted to 40,000. The Romans had achieved a total victory over the Tolostobogii and Trocmi.

7.3.2 The Battle of Ancyra

There was little time to celebrate for the Romans, as an entire new war against the Tectosagi awaited them. As had happened with the Tolostobogii, the Tectosagi and what remained of the Trocmi had entrenched themselves in the mountains, relying on the protection the difficult terrain afforded against the Roman legions. The Tectosagi kept on delaying peace negotiations in an attempt to move their belongings and their women and children to safety across the Halys River. At the final peace negotiations, a Galatian cavalry force ambushed Manlius' small retinue. The consul would have been completely overwhelmed if it was not for a Roman force which happened to be foraging close by and came to his aid, driving off the Galatian mounted troops. The ambush brought an end to the negotiations, sealing the fate of the Tectosagi, and 'the Romans, their hearts on fire with wrath, advanced the next day with their whole strength against the enemy'.[53]

When Manlius arrived at the base of the mountain, he spent two days scouting the area to find the best approaches to the Galatian camp and for any potential ambushes. On the third day, the consul moved to attack the camp, the Galatians responding by drawing up a line of warriors in front of the camp. Manlius divided his army into four columns, taking two columns to attack the Galatian centre, which comprised a large number of Tectosagi and Trocmi. The third column was sent against the Galtian left flank, which was made up of Cappadocians and Paphlagonian allies. The fourth column was sent against the Galatian cavalry, which dismounted due to the terrain being unsuitable for horses and occupied the right flank of the Galatian line.

Livy tells us that the battle was a virtual repeat of the one against the Tolostobogii on Mount Olympus. Manlius first sent his light troops to engage the Galatians. As before, the light troops were supplied with ample ammunition:

> 'When they approached, everything was the same on both sides as in the former battle except courage, which was increased in the victors by reason of their success and diminished in the enemy because, while they themselves had not been conquered, they considered as their own the disaster suffered by men of their own race. And so from like beginnings the affair had the same end. The discharge of a veritable cloud of light missiles overwhelmed the line of the Gauls.'[54]

Appian also highlights the devastating effect of the light troops' missiles upon the compacted ranks of the Galatians:

> 'He enclosed them with his light-armed troops and rode around ordering his men to shoot them at a distance, but not to come in contact with them. The crowd was so dense that no dart missed its mark.'[55]

Livy says that none dared to break from the formation due to the fear of being hit by missiles, which came from all directions. The continuous rain of missiles upon them demoralized the Galatians, and their formation started to disintegrate due to the many casualties and the wounds they suffered. Finally, upon seeing the ranks of the Galatians disorganized and their warriors demoralized, many of them with all manner of missile wounds, Manlius ordered his heavy infantry to charge. The Galatians did not even receive the legionaries' charge before breaking and fleeing to their camp:

> The Gauls, terrified as they were by the recollection of the defeat of the Tolostobogii, carrying missiles fixed in their bodies and wearied both by standing and by wounds, did not endure even the first assault and shout of the Romans.'[56]

The Romans pursued the Galatians to their encampment, hacking them down, but when they reached the camp Manlius had little success in restraining his men from looting. The few men he could command, mostly from the flanks, he ordered to pursue and hunt down the Galatians. Livy says the Galatians suffered 8,000 casualties. The victory brought an end to Galatian resistance and the end of a short but effective war.

7.3.3 Battle Assessment

After his victory, Manlius returned to his winter quarters, where the surviving Galatians sent a delegation to meet with him and sue for peace. The consul and his men grew rich from the plunder they gained from the two captured

Galatian camps, but upon their march home the Romans were ambushed by Thracian tribesmen and much of the booty was lost.

Despite this setback, when he arrived in Rome, Manlius requested a triumph for his victory over the Galatians. However, he was heavily criticized for his unauthorized war and his setback at the hands of the Thracians. He was even mocked for his victory because it was won with 'sling bullets and arrows' and 'no sword was made bloody in combat in the Gallic War'.[57] As mentioned earlier, this goes some way to demonstrate the bias against light troops, especially toward missile units.

Even though his victory over the Galatians was looked upon with disdain, Manlius conducted a very successful and effective campaign. In a very short period of time, he had conquered the Galatian tribes, who were described by ancient authors as some of the most warlike of nations in the East. Livy says that 'of all the peoples who inhabit Asia the Gauls stand first in reputation for war'.[58] Strabo, meanwhile, states that 'the whole race which is now called both "Gallic" and "Galatic" is war-mad, and both high-spirited and quick for battle'.[59] Justinian probably describes their warlike nature the best:

> 'The kings of the east then carried on no wars without a mercenary army of Gauls; nor, if they were driven from their thrones, did they seek protection with any other people than the Gauls. Such indeed was the terror of the Gallic name, and the unvaried good fortune of their arms, that princes thought they could neither maintain their power in security, nor recover it if lost, without the assistance of Gallic valour.'[60]

Manlius fought a war against a fearsome foe on a battlefield that was most unfavourable to his heavy infantry, which was Rome's traditional strength. Yet he clearly understood the value of light troops and recognized their importance for his campaign, as is evident in his efforts to supply them with ample ammunition. Not only did Manlius gain a victory over the tribes, but his success over them can be described as a 'tearless triumph' in the sense that the fatalities among the Romans were so insignificant that there were few who had to weep for the dead. His light troops gained a tremendous victory over the warlike Galatians, one that was so successful and with so few losses on the Roman side that it almost seemed to be without glory.

While the battle might have seemed inglorious to some since it lacked the ideals of *veritas* of melee engagements, it amply demonstrated the need for light troops and just how effective they could be when deployed in the right circumstances. The Galatian advantages of high ground and fortified positions were completely negated by Manlius' effective use of his light infantry.

In the end, despite being begrudged by some, the efforts and effectiveness of the light troops in the war against the Galatians were recognized, and Manlius and his soldiers were awarded a triumph.

7.4 Conclusion on the Battle Assessments

The above-mentioned battles are but three encounters that stress the importance of light troops in ancient warfare. The Battle of Sphacteria demonstrated that the skirmishing hit-and-run tactics of light infantry were highly effective, even against the very best of heavy infantry. It also served as a prime example of the need for combined arms tactics and the need to deploy light infantry alongside heavy infantry as defence against the enemy's light troops. The Battle of Mount Labus brings to light the important role played by light infantry in safeguarding a vulnerable marching column during campaigns and how essential they were in getting an army safely from one point to another. It further highlights the need for light infantry when crossing and fighting in difficult terrain. The Battles of Mount Olympus and Ancyra similarly demonstrate the need for light infantry when fighting on difficult ground. They also go a long way in drawing attention to the effectiveness of range units, particularly if provided with enough ammunition.

The list of important battles where light troops played crucial roles is far from exhausted, and there are numerous more that could be highlighted. Many of these have already been touched upon throughout the various chapters of this book. We have seen how during Demosthenes' invasion of Aetolia, the Athenian hoplites were heavily defeated by the hit-and-run tactics of the Aetolian *psiloi*. In the same battle, Thucydides comments on how the Athenian archers served as an effective defence against the Aetolian light infantry until their arrows ran out. Similarly, the Syracusan light troops wreaked havoc on the Athenian hoplites during the Sicilian disaster, while Iphicrates' peltasts decimated the Spartan heavy infantry at Lechaeum in 391 BC. We also saw the important role played by the Rhodian slingers in the defence of the marching Greek army against the Persian archers. Slingers were successfully used in the armies of Ventidius and Antony against the Parthian horse archers, avoiding a repeat of the disastrous Battle of Carrhae. At the Battle of Himera, the Balearic slingers not only saved the Carthaginian camp from being overrun by the Syracusan forces but used their slings with such devastating effect against the armoured hoplites that they snatched victory from the jaws of defeat.

Chapter 8

Experimental Archaeology

Experimental archaeology as a field of study allows us a unique view of the past through reconstructing ancient artifacts and using them in controlled tests to better understand how they were used. In this chapter, special focus is placed on various arms used by the light infantry, and the effectiveness of these weapons is examined through experimental archaeology. This is done by reconstructing these weapons as faithfully as possible and testing them in various controlled environments.

Javelin and Throwing Strap

8.1 Throwing the Javelin with the Ankylē/Amentum Throwing Strap

The use of a throwing strap (*ankylē* in Greek, *amentum* in Latin) to increase the effectiveness of throwing a javelin dates back as far as the second millennium BC among the Indo-Europeans.[1] These throwing straps were widely attested throughout history, being used by various peoples and cultures. They remained in use well into the Middle Ages and beyond, with their use in Ireland dating as late as the sixteenth century.[2] Throwing straps are well attested in both iconographical evidence, such as the Alexander Mosaic and the Tomb Frieze from Nola, and period literature, incuding Xenophon, Livy and Plutarch.[3]

The widespread use of throwing straps for javelins over such a long period of time is a testament of their effectiveness. Interest in the throwing strap and a desire to test its effectiveness dates back to the Napoleonic era, when tests were performed for the emperor by General Reffye.[4] There are more recent studies on throwing javelins with the *ankylē/amentum* strap, which are also more scientific. These studies have been conducted with regards to how far a javelin's range can be increased by making use of the *ankylē*, and have all examined the javelin thrown in an athletics context, not in a battle scenario. As discussed in Chapter 4 (see 4.4: The Javelinmen and Peltasts in Combat), javelins were thrown from a fairly close range during combat. Accuracy and penetration power was far more important on the battlefield than range.

Previous studies have not examined how the throwing strap increased the kinetic energy of the javelin, compared to without. This section, therefore, examines the 'hitting power' of a javelin by calculating the kinetic energy with which it strikes its intended target when thrown with the aid of a throwing strap, compared to throwing a javelin without.

8.1.1 The Ankylē/Amentum Throwing Strap

Throughout the Graeco-Roman period, the use of a throwing strap to throw javelins was used in hunting game, athletic events and war. There is much iconographical evidence of naked athletes using javelins where the throwing strap is clearly visible.[5] There are also hunting scenes, such as one on a Corinthian vase in the British Museum which clearly shows the hunters throwing javelins with the aid of a throwing strap.[6] It appears that the throwing strap in athletic events was simply wrapped around the shaft as opposed to being firmly fixed to it.[7] In war, however, all evidence suggests that the strap was firmly attached to the shaft (**see plate 33**). Plutarch, for instance, gives an account of Philopoemen being wounded by a javelin that pierced through both his legs. Philopoemen could not remove the javelin due to the throwing strap causing an obstruction:

> 'Here he [Philopoemen] had both his thighs pierced through by a thonged javelin. The wound was not fatal, though severe, and the head of the weapon came out on the other side. At first, then, he was held fast as by a fetter, and was altogether helpless; for the fastening of the thong made it difficult to draw the weapon back through the wound. But since those about him hesitated to attempt this, and since, now that the battle was at its hottest, the ardour of his ambition made him impatient to join in the struggle, by moving his legs backward and forward he broke the shaft of the weapon in two in the middle, and then ordered each fragment to be drawn out separately.'[8]

The javelin lying in the foreground in the Alexander Mosaic, as well as the javelins in the hand of the Samnite warrior in the Nola Tomb, also clearly show the throwing strap firmly attached to the javelin. Even in hunting, the throwing strap seems to be securely attached. The hunting scene on the Corinthian vase mentioned above depicts the Greek hunters chasing down a boar that has already been struck by a javelin. The javelin protruding out of the boar still has the throwing strap attached to it.

Fastening the throwing strap to the shaft as opposed to simply wrapping it around the shaft made a lot of sense when it came to warfare. There was simply no time in a battle scenario for a javelinist to wrap a piece of leather around a javelin, especially if he was trying to avoid the enemy's missile cast at him. It would also take two hands to wrap the strap around the shaft, which meant putting down one's shield and other javelins, wasting precious time in an environment where time was often the difference between life and death.

The throwing strap took the form of a loop, through which the index finger and middle finger of the throwing hand were placed, while the shaft of the javelin rested in the palm of the hand, gripped by the remaining three fingers (**see plate 33**). When thrown, the index and middle fingers continued to pull on the thong after the javelin left the hand. This provided extra leverage, increasing the length of time that acceleration and force were applied to the weapon, similar to the use of a sling when casting a stone. This allowed the javelin to be cast with more accuracy, power and at a further distance.[9]

8.1.2 Method and Materials

An *akōntion* javelin was created for the purpose of experimental archaeology. The javelin head, 16cm in length and weighing 120 grams, was based on an archaeological find from Olynthus (**see plate 34**).[10] Both Theophrastus and Xenophon state that javelins were made out of cornel wood.[11] For the reconstruction I used meranti wood, a hardwood which has a similar density to cornel.[12] Because wood is degradable, no javelin shafts from our period have survived. We thus have to use iconography and literary evidence to determine the measurements of the shaft. We have Polybius' description of the javelin used by the Roman *velites* and Livy's account of the javelin given to the select troops in the war against Hannibal.[13] Polybius states that the *velites'* javelin (he uses the term '*grosphos*' for this particular javelin) shaft measured 2 cubits (96cm), while the head measured an additional span, giving the javelin a total length of approximately 118cm. Livy writes that the select light troops, along with the cavalry, carried javelins measuring 4ft (121cm), not all that different in length compared to Polybius' description. Strabo mentions a Celtic and Germanic javelin with similar dimensions to the *grosphos*.[14] I thus thought it reasonable to make the length of the javelin 120cm. It must, however, be kept in mind that there were a great variety of javelins that existed in the Graeco-Roman world. The all-iron Iberian *soliferreum*, for example, was 1.6–2 metres in length, while the Tomb Frieze of Nola also shows javelins approximately 2 metres long.[15]

The diameter of the shaft of the *akōntion* was based on the 1.7cm diameter of the socket of the Olynthus javelin head. This corresponds well with Polybius' description of the shaft of the *grosphos* as 'a finger's breadth', which is just under 2cm in diameter.[16] The shaft had to be fairly thin for javelinists to carry as many javelins as possible in their non-throwing hand. The fact that the shaft was rather thin is supported by Plutarch's account of Philopoemen having both his thighs pieced by a javelin.[17] As mentioned, Philopoemen broke the javelin shaft in two so that it could be pulled out. For him to have done so supports the thin shaft for javelins. The thinness is partly offset by the fact that high-density hardwood such as cornel was used for javelin shafts, making them relatively heavy for their size.

A circular groove was cut around the shaft of the javelin, approximately 8cm behind the point of balance. This served as a fixing point for the throwing strap and would assist in preventing the leather strap from slipping on the smooth surface of the shaft when pull was applied to the strap.[18] The throwing strap was made from a cut strip of tanned leather 25cm in length and 5mm thick (**see plate 37**).[19]

Range of the Javelin

I acted as the test subject for my experimental archaeology since I had prior experience throwing the javelin.[20] Several javelin throws were conducted without the use of the throwing strap. A maximum of 23 metres was achieved by casting the javelin without the strap. It took a few throws to get used to the throwing strap, and throwing the javelin with the strap initially made little difference. After some practice, however, throwing the javelin with the throwing strap produced some interesting results. A maximum range of 43 metres was achieved with the aid of the strap. The results were replicated with several throws. The throwing strap thus improved my throwing distance by almost 100 per cent.

As mentioned, long-range throws were not the main concern for javelinists, who were more focused on accuracy and penetration power (see 4.4: The Javelinmen and Peltasts in Combat). Even so, the above range experiment did prove that the *ankylē/amentum* increased the force with which the javelin was thrown, and so increased the weapon's hitting power.

To establish just how effective a javelin can be in a combat scenario when thrown with the *ankylē/amentum*, the physics of the javelin in flight and when hitting its target had to be assessed.

The Kinetic Energy of a Javelin Thrown with and without a Throwing Strap

When a javelin is thrown it carries **kinetic energy** while in motion. When the javelin hits its target, it transfers the **kinetic energy** to the target, which is the basic force of impact. **Kinetic energy** (KE) can thus be used to calculate with what impact a javelin strikes its target. **Kinetic energy** is calculated by the formula $KE = \frac{1}{2} mv^2$ (kinetic energy = 0.5 x mass x $(\text{velocity})^2$).

The **mass** of the javelin was established by simply weighing the javelin on an industrial-standard scale; it was calculated to weigh 410 grams (0.410kg). The average **velocity** (Dv) was calculated by the **distance** (s) the javelin was thrown, divided by the **time** (t) it took from release to the moment the javelin struck the target (= s/t). Each throw was conducted over a **distance (s)** of 10 metres, across a smooth flat surface, which was measured by means of an athletics sport measuring tape. Since the angle of trajectory is not steep owing to the short distance and high speed, the effect of the horizontal components could be ignored.

Table 8.1. The kinetic energy with which the javelin was thrown without the throwing strap striking its target.

Javelin cast without throwing strap	KE Equation	Joules generated by throw
Cast 1	$0.5 \times 0.410 \times 15.62^2$	50.01
Cast 2	$0.5 \times 0.410 \times 15.64^2$	50.14
Cast 3	$0.5 \times 0.410 \times 15.63^2$	50.08
Cast 4	$0.5 \times 0.410 \times 15.64^2$	50.14
Cast 5	$0.5 \times 0.410 \times 15.63^2$	50.08

Table 8.2. The kinetic energy with which the javelin was thrown with the throwing strap striking its target.

Javelin cast with throwing strap	KE Equation	Joules generated by throw
Cast 1	$0.5 \times 0.410 \times 21.69^2$	96.44
Cast 2	$0.5 \times 0.410 \times 21.72^2$	96.71
Cast 3	$0.5 \times 0.410 \times 21.73^2$	96.79
Cast 4	$0.5 \times 0.410 \times 21.72^2$	96.71
Cast 5		96.79

The best throw without the aid of the throwing strap took **0.64 seconds** for the javelin to hit the target (see Table 8.1). There was a significant reduction in time when the javelin was thrown with the throwing strap, it taking a mere **0.46 seconds** for the javelin to hit the intended mark (see Table 8.2).

The **velocity** of a javelin thrown without a throwing strap amounted to 15.6 metres per second (Dv = **15.6 m/s**).

The **velocity** of a javelin thrown with a throwing strap amounted to 21.7 metres per second (Dv = **21.7 m/s**).

The **kinetic energy** (measured in **joules/J**) with which the javelin hit the target over a distance of 10 metres could thus be calculated as follows:

Table 8.3. The kinetic energy comparison of javelin upon impact when thrown with and without the throwing strap.

Without *ankylē*	KE = 0.5 x 0.410kg x (15.6m/s)^2	49.9 J
With *ankylē*	KE = 0.5 x 0.410 kg x (21.7m/s)^2	96.5 J

A javelin thrown with the throwing strap thus struck the target with almost double the amount of **kinetic energy**, compared to simply throwing the javelin by hand. The amount of kinetic energy (96.5 J) with which the javelin struck its target over the range of 10 metres is staggering. To put this in perspective, a **kinetic energy** impact of 88 J is considered more than enough to hunt some of the largest, toughest and most dangerous game, such as Cape Buffaloes and Grizzly Bears.[21] Not only does the throwing strap increase the kinetic energy with which the javelin hits its target, but it also increases the weapon's speed and velocity. This would make it much more difficult to evade on the battlefield.

An important find from the experimental archaeology has been demonstrating the armour-piercing capabilities of the javelin when thrown with a throwing strap. In order for a javelin to pierce 1mm-thick bronze, which is the typical thickness of a Classical-era bronze breastplate, the javelin needs a **kinetic energy** of 53.9 J.[22] From this, it became clear that 10 metres or further away, a javelin thrown by hand without the throwing strap was very unlikely to penetrate bronze armour. However, throwing a javelin with the aid of the throwing strap carried more than enough **kinetic energy** to penetrate bronze armour of the period.

Hunting dangerous game such as the Cape Buffalo and Grizzly Bear with a javelin at a distance of 10 metres must be a daunting task, yet there are hunters who demonstrate their skill by doing just that.[23] For light infantry engaging heavy infantry, a distance of 10 metres is an effective fighting range. It was close enough to maximize accuracy and the kinetic energy with which the javelin hit the target, yet gave enough distance for the javelinist to retreat to safety if the heavy infantry charged. There was little to no chance of a heavy infantryman, weighed down in his gear, catching a lightly armed javelinist who had a 10-metre head start on him. The heavy infantry became the most vulnerable targets when they became desperate enough to break from their tight formation to charge the light infantry, such as happened at the Battle of Lechaeum in 391 BC, as we have already seen:

> 'Then the polemarch ordered the first ten year-classes to drive off their assailants. But when they pursued, they caught no one, since they were hoplites pursuing peltasts at the distance of a javelin's cast … when the Lacedaemonians were retiring from the pursuit, being scattered because

> each man had pursued as swiftly as he could, the troops of Iphicrates turned about, and not only did those in front again hurl javelins upon the Lacedaemonians, but also others on the flank, running along to reach their unprotected side.'[24]

Heavy infantry without the support of their own light troops had no answer to these javelinists other than to form up in close formation and wait out the storm of missiles, a futile tactic if a tactic at all.

8.1.3 Conclusion on Throwing the Javelin with the Throwing Strap

Through experimental archaeology, we have demonstrated that the throwing strap increased the effectiveness of the javelin and made it a far more deadly weapon on the battlefield. It significantly increased the force with which the javelin could be thrown. The throwing strap thus increased both the distance a javelin could be thrown and, more importantly within the military context, the kinetic energy with which the javelin struck its target. In our study, a javelin of 0.410kg, thrown using a strap over a distance of 10 metres, struck its intended target with a staggering 96.5 J, while the javelin thrown without a strap struck the target with 50 J. Thrown with a throwing strap, it struck the target with almost double the amount of energy, compared to the same javelin thrown over the same distance by hand. While the javelin thrown without the aid of the strap has very little armour-piercing capabilities, when thrown with the strap, the javelin was indeed able to pierce period armour. Using the throwing strap to throw a javelin made the weapon far deadlier and more effective. There is thus no surprise that the throwing strap saw such widespread use by so many different peoples and cultures, over such a long period of time.

The Sling and Slingshot

8.2 The Lethality and Damage Potential of the Sling

As discussed, the sling brought death on almost every battlefield of the ancient world. Period literature and iconographical sources frequently refer to the sling as a highly effective weapon. Celsus, for instance, notes that a lead sling projectile could travel at such speed and had such a violent impact that it could even penetrate and embed itself into flesh.[25] Onasander speaks of the severe blunt force trauma inflicted upon the body, saying that the 'contusion is mortal without loss of blood'.[26] Diodorus and Cassius Dio also describe the devastating effect these slingshot had on armour.[27]

The purpose of this research study is to examine, through the use of experimental archaeology, the impact and damage a sling could cause on the ancient battlefield and compare the results to the literary texts. The experiment examines both the sling of 'triple' construction, popular among the Greeks, and the 'Balearic' sling of 'single' construction described by Livy. The research also compares the lead projectile popular among the Greeks with the '*mina*' projectile used by the Balearics.

a) Firstly, the kinetic energy of various sling projectiles upon impact was measured. The level of risk of the blunt force injury was then assessed via the Blunt Criterion parameter and converted to the Abbreviated Injury Scale to determine the level of the injury.[28]
b) Secondly, the impact and damage of the lead projectile on period armour was examined upon a reconstructed bronze helmet to test how effective sling projectiles were against the armour of the time.

This study provides us with a better insight and understanding of the use of the sling in war and its effectiveness and lethality as a weapon.

8.2.1 Method and Materials

Slings

A 'Greek' sling was, according to Xenophon's and Livy's description, braided and of triple construction.[29] As mentioned above, the triple construction referred to by Livy is likely a sling in which the cords or arms were constructed separately from the pocket. For the experimental archaeology, I constructed a sling with the cords braided out of hemp, while the pocket was woven out of wool (**see plate 15**). The overall length of the completed sling was approximately 130cm. This was to be used to cast the smaller lead slingshot and the 120g stone (see 'Slingshot Missiles' below).

A second, 'Balearic' sling was plaited out of a single material, and is also known by some as a 'shepherd sling'. This sling was woven completely out of flax (**see plate 16**). It was to be used to cast the large 430g stones (see 'Slingshot Missiles' below).

Slingshot Missiles

Several biconical slingshot projectiles were cast from molten lead, making use of a mould made of industrial putty (**see plate 35**). The slingshots were cast in historical weights (45g) based on several finds from the fourth century BC, which ranged between 25g and 50g (**see plate 36**).[30]

For the *mina* missile, such as those used by the Balearic slingers, water-smoothed stones were selected. I have made the argument that since Diodorus wrote in the first century BC, he made use of the Attic/Euboic standard and thus I selected a stone weighing 430g.

8.2.2 Calculating the Severity of the Injury through Blunt Criterion (BC) and Abbreviated Injury Scale (AIS)

KE as Force of Impact

As discussed in the previous section (see 8.1.2), the impact force of a projectile can be measured in **kinetic energy**. This can be calculated through the equation D (kinetic energy = 0.5 x (mass x velocity)2).

The **mass** of the lead and stone projectiles were calculated by weighing them on industrial-standard electronic scales (see Table 8.4).

Table 8.4. Size of projectiles.

Projectiles	Mass	Diameter
Lead	45g	2cm
Stone	120g	4.13cm
Stone	433g	8cm

The projectiles were thrown at a set range of 15 metres across a flat surface, which was measured using an athletics sports measuring tape.

The average **velocity (Dv)** was calculated by the **distance** thrown **(s)**, divided by the **time (t)** it took from release to the moment the missile struck the target (= **s/t**). As mentioned above, these large *mina*-type stones were thrown at fairly close distances, with an accurate range of 15–20 metres. Each throw was conducted over a **distance (s)** of 15 metres, across a smooth flat surface, which was measured with an athletics sport measuring tape. Since the angle of trajectory is not steep owing to the short distance and high speed, the effect of the horizontal components could be ignored. I have already made the argument that these heavy *mina* stones were thrown over short distances, hence the distance of 15 metres.

In order to establish the **time (Dt)**, several throws over the fixed **distance (15m)** were recorded by video camera. With means of slow-motion tracking, the time between the projectile being released from the pocket of the sling and it striking the target allowed us to establish the **time** for the equation. Again, since the angle of trajectory is not steep owing to the short distance and high speed, the effect of the horizontal components could be ignored.

- The lead projectile took **0.29 seconds** from release from the pocket to striking its target and thus reached a velocity of **55.55 m/s.**
- The stone projectile of 120g took **0.31 seconds** from release from the pocket to striking its target and thus reached a velocity of **48.38 m/s.**
- The stone projectile of 433g took **0.62 seconds** from release from the pocket to striking its target and thus reached a velocity of **26.7 m/s.**

The kinetic energy (measured in **joules /J**) with which the projectiles hit the target over a **distance** of 15m could thus be calculated as follows:

Table 8.5.Lead and stone projectile impact in Jules.

Sling projectile	Formula for Kinetic Energy	KE in Joules
Lead projectile 45g	KE = 0.5 x 0.045kg x $(51.72m/s)^2$	KE =60.14 J
Stone projectile 120g	KE = 0.5 x 0.12kg x $(48.4m/s)^2$	KE =140.44 J
Stone projectile 433g	KE = 0.5 x 0.43kg $(26.78m/s)^2$	KE =154.19 J

The kinetic energy of the stone projectile of 430g upon impact was thus **154.19 J**, while the stone projectile of 120g had a kinetic energy of **140.44 J.** The lead projectile was substantially lower at **60.14 J**, which is still an impressive amount of force. To demonstrate these levels of impact, recent studies have shown that when assessing blunt trauma behind body-armour, residual impact energy values of 70 J for the chest and 25 J for the head serve as threshold values for blunt force trauma.[31]

A very interesting observation was made when the kinetic energy on impact of the 430g stone was compared to that of a smaller 120g stone. I do not consider myself to be an expert slinger, far from it. But I am an experienced slinger, and cast a stone of 120g with ease and reasonable accuracy. However, it took some time to acquire the skill to cast stones of such immense size as the *mina* used by the Balearics. At first, my efforts were lacklustre. After some practice with these huge stones, I was achieving accuracy, but after doing the maths, the kinetic energy with which I hit the target across 15 metres was fairly low at a mere **104 J.** It was much lower than the kinetic energy of **140.44 J** produced by a 120g stone. Even though the *mina* stone of 433g was much larger than the 120g stone, it travelled at a much slower speed when released from the sling.

Two things can be taken from this. The importance of the velocity with which a sling missile hit its target is emphasized. A much smaller stone of 120g far outperformed a stone more than three times its weight in hitting power because of the greater velocity at which it travelled. It is thus understandable why missiles were cast from lead and shaped out of clay to make them as aerodynamic as possible. Secondly, Diodorus' statement about the Balearics who 'in the business of war … hurl much larger stones than do any other slingers' goes some way in demonstrating the skill needed to cast a stone of *mina* size, and that not every slinger was able to cast a stone of such size with efficiency.[32] This further highlights the value of Balearic slingers in war.

With more practice, I was eventually able to improve the speed of my throws, allowing me to hit the target with **154.19 J**. This is not much higher than the 120g stone's **140.44 J**. But there is a big difference between a stone of 120g and one weighing 430g. I can carry a whole bag full of 120g stones, but only a limited number of 430g ones. I can also cast a 120g stone at a much greater distance than a 430g stone. Again, this emphasizes the skill of the Balearics. For them to use the larger stones of 430g means that, because of years of honing their skill, they were able to throw them at much greater velocities than I could and hit their targets with greater kinetic energy. It would take years to learn to effectively throw a stone of a *mina* weight. The Balearic, because of their training and practice, were able to throw these stones with great efficiency, which made them sought-after mercenaries because they hurled 'much larger stones than do any other slingers'.

Blunt Criterion

Blunt Criterion (BC) serves as an energy-based model developed by Sturdivan at the Military Biophysics Lab at Aberdeen Proving Ground, Maryland, to predict injury risk and the severity of blunt force trauma.[33] This parameter is used within the US Department of Defense to assess vulnerability to blunt weapons, projectile impacts and behind-armour-exposures, and thus serves as a good predictor of projectiles thrown from slings. By using Blunt Criterion, we can establish the injury risk assessment of sling projectiles to both the head and the chest. The Blunt Criterion values can then be converted to Abbreviated Injury Scale (AIS) to put the risk of injury into perspective.[34]

Blunt Criterion is defined by the following equation:

$$BC = \ln\left(\frac{\frac{1}{2}MV^2}{W^{1/3}TD}\right) \quad \text{or} \quad BC = \ln\left(\frac{KE}{W^{1/3}TD}\right)$$

- M = **mass** of projectile in kilograms
- V = **velocity** of projectile in metres per second
- W = **mass** of struck target in kilograms
- T = **thickness** of the body wall of target in centimetres
- D = **diameter** of the projectile in centimetres

The mass of the target (W) and the body wall thickness (T) were determined through anthropometric literature and studies.[35] The following values were thus used for the target:

Table 8.6. Values for Target Weight and Thickness.

Target	Weight	Wall Thickness
Human Head	$(W_{head}) = 4.9kg$	$(T_{head}) = 1cm$
Human Thorax	$(W_{body}) = 16.2kg$	$(T_{body}) = 3cm$

The kinetic energy (KE) of the stone and lead projectiles have already been established above, equating to 153.27 J for the stone projectile and 60.14 J for the lead projectile.

The diameter of the projectile as noted in Table 8.6 measured an average of 8cm for the 430g stone, 4.13cm for the 120g stone and 2cm for the lead. It should be noted that previous studies have shown that the diameter of the projectile in relation to the thickness of the target played a significant role when the diameter (D) is more than twice the thickness of the target wall (T).[36] When a small, blunt object hits the body, the area of contact is just the cross-sectional area of the projectile. As the size of the projectile increases, the area of contact is determined by the curvature of the projectile and the body part struck. Sturdivan thus introduced a correction to the effective diameter (D) when it is more than twice the size of the target wall thickness (2 x T). Geometrically, A presents the target area:

$$A = \pi T(D-T);\ D > 2T$$

The effective diameter, D, to be used in the Blunt Criterion equation is thus:

$$D = 2\sqrt{A/\pi} \quad \text{or} \quad D = 2 \times \left(\frac{A}{\pi}\right)^{\frac{1}{2}}$$

In the case of the lead projectile, the diameter is not two times that of the wall thickness of the target and therefore the effective diameter equation does not need to be applied. The stone projectile's diameter is, however, more than twice the wall thickness of the target and therefore the effective diameter equation needs to be applied to the projectile:

Effective diameter of 120g Stone	Effective diameter of 430g Stone
$D = \pi 1(4.13-1); D > 2T$ A = 9.85 $D = 2\sqrt{9.85/\pi}$ or $D = 2 \times \left(\frac{9.85}{\pi}\right)^{\frac{1}{2}}$ D = 3.542	$D = \pi 1(8-1); D > 2T$ A = 21.99 $D = 2\sqrt{25.13/\pi}$ or $D = 2 \times \left(\frac{21.99}{\pi}\right)^{\frac{1}{2}}$ D = 5.29

Table 8.7. Valuwmestone projectile and B) lead projectile impact upon the head and the thorax.

Lead Projectile	120g Stone Projectile	430g Stone Projectile
• M = 0.045kg lead projectile • V = 51.72m/s • W = 4.9kg head / 16.2kg thorax • T = 1cm thickness of combined skull and soft tissue wall • D = 2cm projectile diameter	• M = 0.12kg limestone projectile • V = 48.38m/s • W = 4.9kg head / 16.2kg thorax • T = 1cm thickness of combined skull and soft tissue wall / 3cm thorax combined ribs and soft tissue wall • D = 3.54cm projectile diameter	• M = 0.43kg stone projectile • V = 26.78m/s • W = 4.9kg head / 16.2kg thorax • T = 1cm thickness of combined skull and soft tissue wall / 3cm thorax combined ribs and soft tissue wall • D = 5.29cm projectile diameter

The values for projectile cast at the designated target were as follows:

Table 8.8. BC values of projectile impact on head and thorax.

A) Lead Projectile	
Head	Thorax
$BC = \ln\left(\frac{\frac{1}{2} \times 0.045 \times 51.72^2}{4.9^{1/3} \times 1 \times 2}\right)$	$BC = \ln\left(\frac{\frac{1}{2} \times 0.045 \times 51.72^2}{16.2^{1/3} \times 3 \times 2}\right)$
BC = 2.87	BC = 1.37

B) Stone Projectile 120g	
Head	Thorax
$BC = \ln\left(\frac{\frac{1}{2} \times 0.12 \times 48.38^2}{4.9^{1/3} \times 1 \times 3.54}\right)$	$BC = \ln\left(\frac{\frac{1}{2} \times 0.12 \times 48.38^2}{16.2^{1/3} \times 3 \times 3.54}\right)$
BC = 3.15	BC = 1.65

C) Stone Projectile 430g	
Head	Thorax
$BC = \ln\left(\dfrac{\frac{1}{2} \times 0.43 \times 26.78^2}{4.9^{1/3} \times 1 \times 5.29}\right)$	$BC = \ln\left(\dfrac{\frac{1}{2} \times 0.12 \times 48.38^2}{16.2^{1/3} \times 3 \times 5.29}\right)$
BC = 2.84	BC = 1.34

The impact of a 120g stone projectile thrown from the sling at a **distance** of 15 metres had the highest Blunt Criterion value of **3.2** upon the head, while the lead slingshot had a Blunt Criterion value of **2.87** (Table 7.5). Surprisingly, the large 430g stone had the lowest BC value with **2.84**, just slightly lower than the lead projectile. The 120g stone impact upon the thorax had a BC value of **1.65**, while the lead projectile had a BC value of **1.37**. The 430g stone again had a lower BC value of **1.34**. Even though the *mina* stone of 430g had the highest kinetic energy upon impact, it had the lowest Blunt Criterion value.

Abbreviated Injury Scale (AIS)

The Abbreviated Injury Scale is an anatomical scoring system which is a reasonably accurate way of ranking the severity of injury. Injuries are ranked from 1–6 in severity, with 1 being minor and 6 untreatable and the probability of survival being virtually 0 per cent.[37]

Table 8.9. Ordinal scale used for AIS severity code classification.

AIS	Severity	Type of Injury
1	Minor	
2	Moderate	
3	Serious	Not life-threatening
4	Severe	Life-threatening
5	Critical	Survival Uncertain
6	Virtually Unrevivable	Actually Untreatable

In order to convert BC values to AIS, the equation (**AIS = 1.33 x BC + 0.60**) established by Bir and Viano (2004, 1221) can be used to establish the severity level of the stone and lead slingshot upon target.

Table 8.10 – AIS values of projectile impact on head and thorax

A) Lead Projectile	
Head	**Thorax**
AIS = 1.33 x 2.87 + 0.60	AIS = 1.33 x 1.37 + 0.60
AIS = 4.41	AIS = 2.24

B) Stone Projectile 120g	
Head	**Thorax**
AIS = 1.33 x 3.15 + 0.60	AIS = 1.33 x 1.65 + 0.60
AIS = 4.78	AIS = 2.79

C) Stone Projectile 430g	
Head	**Thorax**
AIS = 1.33 x 2.84 + 0.60	AIS = 1.33 x 1.34 + 0.60
AIS = 4.37	AIS = 2.38

The AIS level of the head impact from the 120g stone projectile was calculated at 4.789 (see Table 8.10). Being above 4.5, the injury level is considered a **Level 5 Critical** injury. The lead projectile has an AIS level of 4.417 and is considered a **Level 4 Severe** injury. The *mina* stone was calculated at 4.37, which is also deemed a **Level 4 Severe** injury. The impact upon the thorax from the 120g stone produced an AIS **Level 3 Serious** injury, while the lead projectile and 430g stone both produced **Level 2 Moderate** injuries.

To further put these levels of injuries in context, a study on the mortality risk of AIS levels has demonstrated that an AIS Level 5 Critical single injury has a 24.5 per cent mortality rate, while a Level 4 Severe injury had a 9.1 per cent mortality rate.[38] Level 3 and Level 2 had substantially lower rates of mortality at 1.8 per cent and 0.9 per cent respectively. These percentages are based on modern medical treatment, and the mortality rate would undoubtedly have been higher in ancient times considering the rudimentary medical treatment of the time.

8.2.3 Projectile Impact upon Armour

The anti-armour potential of slings is abundantly attested by the ancient authors, as noted above. In order to test the effectiveness of a lead projectile cast from a sling, a replica bronze helmet of the Corinthian type was used as the designated target (**see plate 19**).[39] The helmet thickness was between 1.2mm and 1.5mm, which was typical of helmets from the time.[40] The helmet was placed on a pole with a 3cm diameter, which was loosely planted in the ground so that it could easily give under force. A 45g lead projectile was cast at the helmet from a distance of 15m. The lead projectile impacted at the centre of the forehead just above the brow. The point of impact where the lead projectile hit the metal was pushed 1.4cm towards the back. The impact was severe and left such a dent that the helmet was bent out of shape and it was no longer possible to fit over the head (**see plate 20**).

While the helmet took the brunt of the force, the wearer would have undoubtedly taken a severe blow to the head from the impact. The blow would probably not have killed the wearer, but would surely have stunned and incapacitated him to some extent. This would have made him extremely vulnerable and left him exposed to other dangers on the battlefield. This force of impact would have probably caused a deep laceration, with blood streaming across his face. According to Livy, Lucius Aemilius Paullus, one of the Roman commanders at Cannae (216 BC), 'received a severe wound from a sling at the very outset of the battle' yet managed to still issue orders to his men.[41] This 'severe wound', however, eventually weakened the consul, to the extent that he was no longer able to control his horse and had to dismount. It is unlikely that Paullus was struck in any other place than the head for him to have eventually been too weak to ride a horse. The consul would have worn a helmet, through which it is probable the sling projectile struck him. The experiment demonstrated with what severe force Paullus could have been hit, even by a small lead projectile of 45g and through the protective headgear of the day.

This type of impact upon the helmet mirrors the other ancient literary sources, such as Cassius Dio, who wrote that 'they inflicted severe injury upon all, even upon the men in armour', and the account of the prophet Mohammed sustaining such a blow from a sling that it caved in the cheek-piece of his helmet and drove it into his face.

While the financial limitations of the project allowed for only one helmet to be tested, and since the integrity of the helmet was already compromised by the impact from the light 45g lead missile, the heavier 120g and 430g stone projectiles could not be tested. It is, however, reasonable to conclude that since the 120g stone projectile had an impact force of 140.44 J, more than double that of the 45g lead projectile, the stone would have caused even more damage to the armour and its bearer. The 430g stone also had a higher KE upon impact, but as we have seen, the BC and AIS values were lower. In the hands of a skilled Balearic slinger, however, there is little doubt that the BC and AIS values would be much higher. Both the 45g lead and 120g stone projectiles were dwarfed when compared to the *mina* (430g) projectiles used by the Balearics. While I was unable to produce effective results with the *mina* stone, the elite Balearics slingers of the time would have the skills necessary to cast these large stones with much greater effect. The Battle of Hiemera, where the armour of the Greek hoplites was shattered by the slingers, is really put into context if one looks at the damage done to the bronze helmet by a light projectile of a mere 45g.

8.2.4 Conclusion on the Sling

The sling was a devastating range weapon that was cheap and easy to construct and required little skill to wield. It was held a range advantage over other missile weapons and was highly effective against armour, even a small sling bullet of 45 grams. It is consequently no wonder that Onasander states that 'the sling is the most deadly weapon used by the light troops'.[42]

We have also seen that the Balearic slingers had the ability to 'hurl much larger stones than do any other slingers', weighing a *mina*. The experimental archaeology has demonstrated that it took skill and training to become effective at throwing such large stones, much more so than lead and stone slingshot between 45 grams and 120 grams. It was this skill of casting these large stones effectively that set the Balearics apart and made them such sought-after mercenaries.

The Bow and Arrow

8.3 Arrowheads: The Cretan Archer vs the Persian Bowman

As we saw in Chapter 2, the Persian archers in the *Anabasis* outranged their Cretan counterparts:

> 'And the Greek rear-guard, while suffering severely, could not retaliate at all; for the Cretan bowmen not only had a shorter range than the Persians, but besides, since they had no armour, they were shut in within the lines of the hoplites.'[43]

Many scholars have made the assumption that the Cretans used lighter/weaker bows or that they used the inferior self-bow. We do, however, see a few passages later that the Cretans were able to make use of Persian arrows, and with some practice fired them to great effect at long range:

> 'The Persian bows are also large, and consequently the Cretans could make good use of all the arrows that fell into their hands; in fact, they were continually using the enemy's arrows, and practised themselves in long-range work by shooting them into the air.'[44]

The passage also tells us that the Persians and Cretans used bows that were very similar in strength or draw weight, with it being said that 'the Persian bows are also large' like those of the Cretans. I have already pointed out in Chapter 2 that the arrows used in traditional archery were tailor-made for their bows. Shooting arrows with the incorrect spine or where the arrow was too light would cause damage to the composite bow. For the Cretans to shoot Persian arrows, the bows had to be of a very similar construction and draw weight. I made the argument that the Cretans used heavier arrowheads than the Persians and that this was in all likelihood the reason why the Persians outranged the Cretans. Once they were able to use 'Persian' arrows, the Cretans – along with the Rhodian slingers – were more than able to hold their own against the Persians:

> 'But when the Rhodian slingers and the bowmen, posted at intervals here and there, sent back an answering volley, and not a man among them missed his mark (for even if he had been very eager to do so, it would not have been easy), then Tissaphernes withdrew out of range with all speed, and the other battalions followed his example.'[45]

Whereas the Greek archers were used to sniping at their individual battlefield targets at close range, they now shot parabolic shots 'in long-range work by shooting them into the air' at a mass of Persian soldiers. Because there were so many soldiers, 'not a man among them missed his mark'.

One of the most popular and largest types of Greek arrowheads is the Type IA 5, according to Baitinger's typology.[46] It is distinguished by its greater width, weight and the two barbs that protrude from the back of the arrowhead parallel to its tang. It has a strong connection with Crete and is commonly called the 'Cretan type'. This is very much because such arrowheads feature on coins minted on the island of Crete. These arrowheads, however, appear widely across the Mediterranean and beyond, their use dating back to Mycenaean times, and it is impossible to reliably determine its origins. Even though it cannot be established as having Cretan origins, the arrowheads were certainly used by Cretan archers, as established by the Cretan coins and these types of heads having been excavated on the island and in areas associated with Cretan mercenary archers. They are most commonly cast in bronze to facilitate mass production, but iron heads of this type do appear in the archaeological record.[47] They are almost always tanged as opposed to socketed, even the iron versions.

The two most common Persian arrowheads were the smaller trilobate and the leaf-shaped, two-bladed versions. Both are believed to have been introduced to the Near East by the Scythians. Examples of both have been found at battle sites such as Marathon and Thermopylae, as well as many other sites connected with Achaemenid Persia. As with the 'Cretan' arrowhead, the 'Scythian' arrowheads are commonly cast in bronze, although several examples of iron do exist. The vast majority of the arrowheads are socketed as opposed to tanged. One theory is that arrow shafts were of reed, and being hollow were more suited to socketed heads as opposed to tanged. These heads are fairly small, weighing between 2 and 8 grams, with the majority under 5 grams. It must be noted that these 'Scythian' arrowheads were popular throughout the Mediterranean and were also used by Greek archers.

This section examines the range advantage the Persian archers enjoyed over their Cretan counterparts in Xenophon's *Anabasis*. It further examines why one would favour larger arrowheads over smaller, lighter heads.

Because it is impossible to determine the exact strength of ancient bows, the material of the arrow shafts and the weather conditions (did the Persians shoot downwind, for instance?), the experiment revolves around hypothetical scenarios, making it more of an enjoyable exploration rather than a precise scientific endeavour.

8.3.1 Materials

Composite Bow

As established, both the Persians and Cretans used the composite bow, and they were of a similar draw weight and size if we follow Xenophon's account in the *Anabasis*. Csaba Grózer from Grózer Archery made a replica of a period horn bow based on the Scythian design (**see plate 1**). The horn bow draws 75lb (34kg) at 28in (71.12cm) and approximately 80lb (36kg) at 30in (76.2cm). The draw weight of the bow falls within what military historians deem as military-weight bows, being above 70lb. As we saw, it is very difficult to determine the draw weight of ancient bows due to the lack of archaeological finds. The few finds that we do have support the chosen draw weight of 80lb. The Scythian bow discovered at Xinjiang had a minimum estimated draw weight of 80lb.[48] The first and second-century AD Parthian bow found at Yrzi is also believed to have had a draw weight of approximately 70lb.[49] The reconstruction of the early Pontic Scythian find from Vishnevka had a draw weight of 55lb at 23in, and thus approximately 79lb at 28in.[50] Another replica of a very fragmented Scythian-style bow discovered at Aržan drew approximately 72lb at 23in (97lb at 28in), but broke under stress tests, suggesting that the original bow was of a lower draw weight.[51] It must be remembered that the draw weight of these reproductions are estimates, the difficulty to determine the draw weight through reproduction being highlighted by Karpowicz:

> 'It must be emphasized, the draw weight must be taken here as only an indication of the possible range for these bows. Very small variations in

dimensions will result in substantial changes in weight. For example, a bow only 0.1cm thinner will be 15lb lighter, 0.2cm thinner about 30lb lighter. Narrowing the bow by 0.1cm will bring about [a] 7lb drop in weight, by 0.2cm [a] 15lb decrease. This could be indeed the case, if the sinew layer was thicker than the estimated 0.1cm. A 2cm longer bow will be about 5lb lighter as well. Moreover, the cross-sectional and side profile of the bows could vary between different specimens, as well as the flexibility of the outer sections of bows, which came to be nearly rigid in the replica [of the Xinjiang find]. These, together with the physical degradation in use and with the changes of ambient humidity and temperature, do not allow for predictions greater than ± 30lb. It would be reasonable to estimate the draw weight range of Scythian bows to be from 80lb to 140lb, which falls close to a similar estimate for other composites.'[52]

As discussed, military historians and ancient texts estimate the draw weight of horse bows to be in the range of 80lb due, to the fact that horse archers could only use their upper body to draw the bow.[53] I have already mentioned that an overwhelming amount of iconographical evidence depicts Graeco-Roman archers as kneeling while shooting their bows. For a similar reason as horse archers, shooting from a kneeling position would require a lighter bow. From practical experience, I find that I cannot shoot bows much heavier than 80lb comfortably and accurately in the kneeling position, which further supports the use of 75–80lb reconstruction.

Arrowheads and Shafts

For the experiment, a bronze Type IA 5 'Cretan' replica arrowhead was cast by a bronzesmith (**see plate 31**). The Cretan arrowhead reproduction was based on those discovered at Jebel Khalid and those excavated at the Stymphalos Sanctuary. The most well-preserved of these finds measured between 8cm and 8.6cm and was 1.6–1.8cm wide at its broadest, weighing between 12 and 14 grams. Larger and heavier versions existed; several of the largest heads excavated at Stymphalos, which weighed 14 grams, were missing tips, barbs and tangs, which means they would have been even heavier if they were undamaged. The bronze replica measured 8.2cm in length, 1.8cm in width and weighed 15 grams.

I used a bilobate iron arrowhead similar to the larger 'Scythian' arrowheads used by the Persians (**see plate 31**). The arrowhead weighed 7 grams, placing it at the heavier end of Scythian arrowheads (the heaviest weighing 8 grams). Both arrowheads were mounted on wooden shafts weighing 35 grams (550 grains) each.

8.3.2 Method and Results

Once again, I acted as the test subject of the experiment. I am an experienced traditional archer with more than 14 years of experience. I am an avid war bow archer and shoot bows well over 90lb in draw weight.

The experiment was conducted on a wide, open beach under calm weather conditions. Both arrows were shot underhand at the same elevation in a parabolic arch. The distance was measured with a retractable measuring tape (**see plate 38**). The distance of each shot is shown in Table 8.11 and Table 8.12 below.

Table 8.11. Arrow range of 'Cretan' and 'Scythian' arrowheads.

Range of the 'Cretan' arrowhead	Range of the 'Scythian' arrowhead
121 metres	162 metres
125 metres	163 metres
123 metres	162 metres
123 metres	160 metres
121 metres	163 metres

Table 8.12. Arrow Range of 'Cretan' and 'Scythian' Arrowhead.

Average range of the 'Cretan' arrowhead	Average range of the 'Scythian' arrowhead
122 metres	162 metres

From the tables above, we can see that the arrows with the smaller 'Scythian' arrowhead, as expected, enjoyed an average of 40 metres range advantage over those with the larger 'Cretan' head. This is taking into consideration that for the experiment the heavier 7 gram 'Scythian' arrowhead was used and that that the range would be even further if the lighter 2–5 gram heads were used.

But if the smaller 'Scythian' head allowed for a greater range advantage, why would the Cretans and other archers favour the larger Type IA 5, especially if the larger arrowhead was more expensive to construct? The answer lies in the different fighting styles. The Persians were able to field large numbers of archers who would rely on shooting vast amounts of arrows in a parabolic arch at enemy formations, as opposed to individual soldiers. The smaller 'Scythian' arrowhead was more fitting for the longrange volleys that 'blot out the sun'. Using smaller arrowheads meant a longer range, allowing the larger number of archers to drown their enemy in arrows from further away. Due to their reliance on shooting high volumes of arrows at a mass of soldiers, the Persians also found it economically advantageous to make use of smaller arrowheads consisting of less metal.

Cretans were deployed in much smaller numbers, and as mentioned before relied on sniping at individual targets on the battlefield, shooting at fairly close range to maximize the effects of every arrow. Because of shooting at close range the gravitational effect on the arrow was minimal. Using large arrowheads had great advantages over the smaller ones. The larger arrowhead automatically made the arrow heavier, which allowed for better penetration. The larger arrowhead also made for a larger, deeper and more devastating wound, which was far more severe than the wound created by the smaller arrowhead. As mentioned, archers often had to target exposed limbs such as arms and legs. This meant that arrow wounds were often non-fatal, but the wound only needed to be severe enough to debilitate an enemy. The larger wound from the 'Cretan' arrow was far more likely to force a soldier from the field of battle, incapacitate him, weaken him or make him vulnerable to other types of enemy attacks than the smaller wound from the Persian arrowhead. The smaller Persian arrowhead also generally lacked barbs, making it easier to remove. In contrast, the broad 'Cretan' head with its large barbs made it nearly impossible to pull out, requiring advanced medical assistance to remove.

I have also mentioned that Cretans were highly effective against cavalry because of their use of large arrowheads. This is reflected in modern bowhunting with the use of 'broadheads', not unlike the 'Cretan' arrowhead, for hunting, which allows arrows to penetrate deeper, passing through the animal, severing arteries, and cutting through vital organs.

8.3.3 Conclusion on the Arrowheads

It made sense that Cretans, being individual warriors on the battlefield who fought at short range, would favour a much larger and heavier arrowhead such as the 'Cretan' over the smaller 'Scythian' arrowhead. This would explain why they were at first outranged by the Persians in Xenophon's account of the *Anabasis*. However, as soon as they were able to obtain Persian arrows and spent some time practising long-range parabolic shots with these, they were able to match the Persian archers.

Chapter 9

Conclusion

Although the heavy infantry took centre stage in the major pitched battles of ancient warfare, this was just a small (albeit significant) part of the overall campaigns. Ancient military campaigns consisted of numerous long marches through difficult terrain (including gruelling forced marches), countless skirmishes, tactical operations over broken ground, the capture of strategic positions, scouting and reconnoitring. In these operations, the light infantry played a crucial role in ensuring success. While the heavy infantry may have received the glory and fame in the large-scale battles, their achievements relied heavily on the foundation created throughout the campaign by the actions of the light troops. Despite being overlooked or undervalued by ancient authors, the light infantry – the unsung heroes of Graeco-Roman warfare – bore the heaviest burden of the campaign, playing a vital role that cannot be denied.

On the March

For an army to reach its destination, it had to form a marching column. These columns marched immense distances over mountains, through deserts and forests, and across rivers. Lee calculated that Xenophon's army of the *Anabasis* consisting of 10,400 hoplites marching four abreast, would have stretched 5.5km.[1] This was a fairly moderate-sized army and was dwarfed by later Hellenistic armies, such as those of the Seleucids and Ptolemies, who could field hosts of 70,000–80,000 fighting men, excluding the baggage train and non-combatants. There would be open plains where the column could spread out and march more men abreast, but through the mountain passes, like those of the Cilician Gates, the column would be reduced to only a few men abreast. Even on a flat plain, the column would move fairly slowly. Progress would be far slower when it snaked through difficult broken terrain or during bad weather. Xenophon's army, marching back to Greece, averaged about five *parasangs* (approximately 24–28km, or 15–17 miles) a day, with roughly six hours of marching every day.[2] Engels calculated that Alexander's Macedonian army also averaged about 24km a day. Vegetius states that a Roman soldier of the fourth century AD was expected to be able to march 32km (20 miles) a day, and had to complete the distance over a time of five hours.[3]

During the march, the column was extremely vulnerable to attack. The light troops and cavalry were heavily relied upon to protect the baggage train and the heavy infantry, who were unable to operate outside of their fighting formation. Vegetius specifically mentions the severe dangers a marching column faced while on campaign, highlighting that the march was even more perilous than actual battle:

> 'Those who have made a careful study of the art of war assert that more dangers tend to arise on the march than in battle itself. For in battle everyone is armed, and they see the enemy at close quarters and come mentally prepared for fighting. On the march, the soldier is less armed and less alert; he is thrown into instant confusion by a sudden attack or concealed ambush. Therefore the general should take steps with all caution and prudence to ensure that the army suffer no attack on the march, or may easily repel a raid without loss.'[4]

Vegetius places special emphasis on the important role played by light infantry and cavalry for the protection of the marching column. Xenophon's army, as we have seen, suffered tremendously from the continuous attacks made by the Persians upon the marching Greeks. That was until Xenophon sufficiently rearmed many of his troops as light infantry, particularly slingers, to ward off the Persian range troops. We also see that Xenophon's column while on the march had as its vanguard the light infantry to serve as defence for the heavily armoured hoplites. The rear had only hoplites and no light infantry, as there was believed to be no danger here, emphasizing the importance of light infantry as the primary means of defence on the march:

> 'Here Cheirisophus, with his own division and all the light-armed troops, led the van, while Xenophon followed behind with the hoplites of the rear-guard, but without any light troops at all; for there seemed to be no danger of any pursuit from behind while they were proceeding uphill.'[5]

During Demosthenes' campaign in Aetolia, it was the archers who served as the only defence against the Aetolian javelinists when the army was caught in an ambush. The Syracusan disaster saw the retreating Athenian column, without the sufficient protection of light troops, being picked apart by the enemy light infantry. Antiochus's gruelling march through the Elbruz Pass was only possible because he had an effective force of light infantry protecting the vulnerable column. Philip V relied on his elite Hellenistic *peltasts* and other light infantry, especially Cretans and Illyrians, in conjunction with his cavalry to defend the marching column during his numerous campaigns. Philopoemen has the majority of his light infantry and cavalry protecting the rear of his marching column because he believed the Spartans were planning to attack them from behind.[6] Sallust describes the Roman marching column moving through enemy territory of the Numidians sufficiently protected by light infantry, such as bowmen and slingers.[7]

There were also forced marches where an army had to get from one point to another in as short a time as possible. The equipment of the heavy infantry tended to impede their movement speed, so generals relied on their light infantry to make these gruelling fast-paced marches. Ancient literature is full of examples of generals leaving behind the slow-moving heavy infantry and taking with them only the light infantry and cavalry. Antiochus the Great's forced march to relieve his besieged ally, Theodotus, at Ptolemais and his forced march to capture the Arius River crossing serve as prime examples.[8] Similarly, Marchus Claudius force-marched to relieve the city of Clastidium, besieged by the Insubres, taking with him only his light infantry and cavalry.[9]

During these forced marches, the light troops played an undeniably vital role in safeguarding the marching column, and were the most suited for when rapid marches were required.

Operating in Difficult Terrain

Intelligence gathering is a vital component of any period of warfare. This was again a role for the light troops. A marching army would inevitably have to march through mountain passes, thick forests and other areas of difficult terrain, where the marching column would be at its most vulnerable as these places were ideal for ambushes. These areas had to be thoroughly scouted to ensure the column could pass safely through. In the more open areas, light cavalry could perform scouting duties, but even then they would often be accompanied by light infantry as support units. However, many of these mountain passes with their extreme outcrops, or areas with thick overgrowth, were inaccessible to cavalry and the scouting could only be performed by the light infantry. Their light equipment made them most suited for carrying out scouting duties in broken terrain and collecting information on enemy positions and forces. Once again, Antiochus' crossing of the Elbruz Pass serves as an example of the light infantry scouting the difficult terrain and locating the enemy's ambushes. Antiochus also used his light infantry (*euzōnoi*) prior to scouting the mountainous terrain and the enemy forces of the Ptolemies who had occupied the Porphyrion Pass.[10]

They were obviously ideally suited to move quickly to capture and defend strategic positions such as hills, passes and river fords. Light infantry and elite *peltasts*, in conjunction with cavalry, were used by Antiochus to capture a river crossing.[11] Philopoemen used his light *caetrati* to defend a stream serving as a water source for the Achaean army.[12] During the war between Caesar and Senatorial forces in Spain, Africanius sent his Spanish *caetrati* (*caetratorum cohortis*) light infantry to secure a strategic hill that would allow him and his army a superior defensive position against Caesar's forces.[13]

For the same reasons that they were ideally suited for scouting and capturing strategic positions in difficult terrain, light troops served as the main offensive arm, and often the only capable force, of ancient armies when combat took place in broken terrain. In the words of Onasader:

> 'If the battle should happen to be in country that is level in some places but hilly in others, then the light-armed troops [ψιλοί] should by all means be stationed in the uneven section, and then, if the general himself should have seized the plain and some part of the enemy's phalanx should possess the heights, he should send against them the light-armed troops [ψιλοί]; for from the uneven ground they can more easily hurl their weapons and retreat, or they can very easily charge up the slopes, if they are agile.'[14]

Fighting in broken terrain would have severe restrictions upon the cavalry and impede the deployment of the heavy infantry by breaking up their formations. Because light infantry was unencumbered by heavy war gear for their style of fighting, and fought as individual combatants, not relying on formations, they suffered far less in broken terrain. Philip V left behind his cavalry and heavy infantry to advance on the forces at Thalamae because the 'approaches to it are narrow and difficult and the place itself secluded and not easily entered'.[15] I have already mentioned Antiochus using his light troops as scouts in the Porphyrion Pass. When the king found the Ptolemaic fortified position, he also used his light infantry to dislodge the enemy troops and force his way through the pass. During the Second Punic War, Hannibal Barca, while escaping entrapment in a valley in Campania, sent his *lonchophoroi* and Spanish light infantry against a Roman force that occupied a hill.[16] Similarly, the Romans used their light infantry to capture a hill occupied by Hannibal's light infantry, who had been sent to seize the strategic position overlooking the Roman camp at a later stage of the war.[17] Not unlike the Roman campaign against the Galatians (discussed in greater detail above), the Battle of Baecula (209 BC) was won by the Roman light troops, who were able to scale a hill and fight in the difficult terrain. At the Battle of Axona (57 BC), Caesar used his light troops (Cretans, Balearics and Numidians) to great effect against a Gallic army whose formation was in disarray after crossing a marsh:

> 'Our men, attacking in the river the disordered enemy, slew a great part of them. By the immense number of their missiles they drove back the rest, who, in a most courageous manner were attempting to pass over their bodies, and surrounded with their cavalry, and cut to pieces those who had first crossed the river.'[18]

Philopoemen, the great Achaean commander, had to advance his army to confront Spartan forces who had occupied a much more favourable position. In order to do so, he had to move his army through uneven terrain, which Livy describes as 'rough country'. Livy noted that the only way Philopoemen's army could advance through the difficult terrain was if it was screened by light infantry.[19]

The examples of the Elbruz Pass and Rome's campaign against the Galatians, which have been discussed in detail elsewhere, serve as other excellent examples of light infantry combat in broken and difficult terrain.

Range Combat and a Counter to Heavy Infantry

The typical weapons associated with the light infantry were range weapons: the bow, javelin and sling. Being able to fight from a distance allowed light troops to avoid the dangers of hand-to-hand combat, which in turn meant that they did not have to be clad in heavy armour or carry heavy shields for protection. They thereby retained their nimbleness and dexterity which allowed them to operate in difficult terrain. This also made them an effective force against heavy infantry. They were able to engage heavy infantry at range, and because of their light arms could easily withdraw if the heavily armed enemy attempted to close with them. Heavy infantry were simply too encumbered by their gear to catch light troops. Heavy infantry who had no or insufficient numbers of light infantry to counter those of the enemy would find themselves in a very vulnerable position. The fast-moving light infantry could easily outflank their enemies and snipe at them from a variety of angles. The heavy infantry could only close ranks and try to weather the storm of missiles. Breaking ranks in any attempt to catch the light troops only exposed them to more of the enemy's missiles. Sparta's over-reliance on their 'superior' hoplites saw them caught in this particular scenario on more than one occasion. The Battle of Sphacteria, discussed in detail above, was Sparta's greatest's defeat suffered during the Peloponnesian War, its hoplites having no means to defend against the hit-and-run tactics of the Athenian light infantry. The pro-democratic leader Thrasybulus, in his war against the Oligarchy of the Thirty Tyrants, used his light troops to great effect against the Spartan hoplites sent to reinforce the Tyrants, who once again lacked adequate light infantry of their own:

'And the light troops, rushing forth at once, set to throwing javelins, hurling stones, shooting arrows, and discharging slings; then the Lacedaemonians, since many of them were being wounded and they were hard pressed, gave ground, though still facing the enemy; and at this the latter attacked much more vigorously.'[20]

At the Battle of Lechaeum (391 BC), Sparta's heavy infantry hoplites, who were 'unaccompanied by either peltasts or cavalry', were decimated by the light troops of Iphicrates:

'[T]he peltasts [retired] before the hoplites got near them; and further, when the Lacedaemonians were retiring from the pursuit, being scattered because each man had pursued as swiftly as he could, the troops of Iphicrates turned about, and not only did those in front again hurl javelins upon the Lacedaemonians, but also others on the flank, running along to reach their unprotected side ... the Lacedaemonians themselves kept continually becoming fewer and fainter of heart, while their enemies were becoming bolder, and those who attacked them continually more numerous.'[21]

Cassius Dio paints a similar picture of Sextus Pompey's light infantry attacking the heavy infantry of Cornificius, a general of Octavian:

'Both cavalry and light troops (ψιλοί) attacked him [Cornificius] from a distance, not daring to come to close quarters, and proved exceedingly troublesome to him; for they would not only attack whenever opportunity offered but would also quickly retreat again, whereas his men, being heavy-armed, could not pursue them in any case owing to the weight of their armour ... Consequently, they were suffering many injuries and could inflict none in return; for, in case they made a rush upon any of them, they would put them to flight, to be sure, but being unable to carry their pursuit to the end, they would find themselves in a worse plight during their retreat, since by their sortie they would become isolated.'[22]

Nimble, fast-moving light infantry, impossible to catch, proved deadly against heavy infantry.

Siege Warfare

We have also seen how being able to fight at long range gave light infantry a special place in siege warfare. They were the first to engage the enemy, whether defending the walls or attacking them. In the case of besieging a city, they were essential in using their range weapons to clear the battlements of enemy combatants. If being besieged, their role was to keep the besiegers from reaching the walls. Incendiary arrows and javelins, and even heated clay slingshot, had the potential to set aflame buildings and fortifications. From the defenders' point of view, they could be used to set alight enemy siege craft.

Ambushes and Surprise Attacks

Ambushes required troops to hide amongst trees, boulders and hills, areas that by nature would be rough terrain. It also necessitated quick decisive action to launch a surprise attack. Being heavily armoured and dependent on their well-ordered formation to fight made heavy infantry far from ideal at ambuscades. It is thus not surprising that when ambushes and surprise attacks are described in the literature, it is almost always performed by the light infantry.

Thucydides relates how the Amphilochian light infantry successfully laid ambushes for the heavily armoured Ambraciot hoplites:

> 'But the paths were beset; the Amphilochians were lightly-armed, and in their own country which they knew, while their enemies were heavy-armed and the country was strange to them. And so, not knowing which way to turn, they fell into ravines and into ambuscades which had been set for them, and perished … the Ambraciots were cut to pieces, and but few out of many returned home to their city.'[23]

Xenophon used his light infantry to set an ambush in the mountains for the pursuing Persians:

> '"Well," said Xenophon, "I am ready to set out with the rear-guard, as soon as we have dined, to seize possession of the mountain. And I have guides, too; for the light troops set an ambush and captured some of the stealing rascals who are following us."'[24]

Antigonus, under the cover of darkness, also defeated Cleitus with the use of his light troops.

Philip V staged a successful ambush against the Aetolians, hiding his elite *peltasts* and Illyrian light infantry in the hills. When the Aetolians attacked his marching column's rear, the ambush was sprung and the attacking Aetolians were caught in the trap, suffering heavy losses.[25] In his war against the Illyrians, Philip used his *peltasts* to lure the enemy out of their fortifications and into an ambush set by his light infantry.[26]

Philopoemen used a devastating night attack against the Spartan forces of Nabis with his *caetrati* light troops:

> 'He collected small boats in a secret haven in Argive territory; into them he loaded lightly-equipped soldiers, mostly *caetrati*, with slings and darts and other kinds of light ordnance. Then, skirting the shore, when he came to a headland near the camp of the enemy, he landed and travelling over familiar trails by night came to Pleiae and, the sentinels being asleep, like men in no immediate peril, hurled firebrands upon the huts on every side of the camp. Many were consumed by the flames before they knew of the enemy's approach, and those who did know of it were able to bring them no aid. With sword and fire everything was destroyed; a very few escaped from this two-fold destruction to Gytheum and the larger camp.'[27]

The light infantry were far more apt at hiding in the undergrowth or in the hills and could move much faster and were far more nimble, making them ideally suited for surprise attacks. While ambushes could and were at times performed with well-hidden heavy infantry, these and surprise attacks were undoubtedly the realm of the light infantry.

Support Troops

Light infantry often served as an effective support unit. This is particularly true for cavalry. Because they were unencumbered by their gear, light infantry were swift enough to act in support of mounted troops. Just prior to the Battle of Cannae (216 BC), Hannibal Barca attacked the Roman marching column with his cavalry, supported by light infantry.[28] The Achaeans raided the Aetolian territory with their cavalry and light infantry.[29] Prior to the Battle of Ilipa (206 BC), both the Romans and Carthaginians sent out their cavalry, supported by light infantry, to skirmish against one another.[30] In the events leading up to the Battle of Cynoscephalae (197 BC), the cavalry of both Philip V and Flaminius, supported by light troops, became engaged in a heated skirmish when they ran into one another during a reconnaissance mission. We also see that Caesar used light infantry to supplement his cavalry as a countermeasure to Pompey's numerically superior cavalry.[31] This was also the case for pitched battles. There are numerous examples of open-field battles, especially during the Hellenistic era, with light troops acting as support units for cavalry. At the Battle of Magnesia (190 BC), Antiochus the Great ordered a cavalry charge, supported by auxiliary light infantry, at the Roman left flank.[32] At the same battle, the Roman right wing, commanded by Eumenes of Pergamum, charged the Seleucid left. In an interesting reversal, the latter charge was made with missile-armed light infantry, supported by cavalry.[33] Xenophon, in his *On the Cavalry Commander*, specifically mentions the disadvantage of horsemen without the support of light infantry when faced with cavalry who have light infantry in support:

> 'Another duty of a cavalry commander is to demonstrate to the city the weakness of cavalry destitute of infantry as compared with cavalry that has infantry attached to it.'[34]
>
> [...]
>
> 'Both divisions [of cavalry] should have an infantry contingent; and if the infantry, hidden away behind the cavalry, came out suddenly and went for the enemy, I think they would prove an important factor in making the victory more decisive; for I have noticed that a surprise cheers men up if it is pleasant, but stuns them if it is alarming.'[35]

Pitched Battle

Throughout the ancient world, it was common practice for the light troops to commence the battle. They formed a screening force in front of the heavy troops, protecting them from the enemy light troops. This particular role of light infantry was a very important one. We have seen how deadly light missile infantry could be against heavy infantry who had insufficient or no light infantry of their own. If the heavy infantry did not have enough light troops to protect them, the enemy's light troops could harass their lines and weaken their formations before the lines of heavy infantry clashed. When the heavy infantry met, the light troops, because they were nimble and fast-moving, could outflank the enemy's heavily armed forces and rain down missiles upon their exposed flanks. It was thus essential for armies during pitched battles to have sufficient light infantry to defend against the opposing light troops. We see that light infantry took up position on the flank of the heavy infantry, acting as a shield to defend their vulnerable flanks from possible light infantry attacks.

The concept of deploying light troops in front of the heavy infantry before the start of pitched battles served a purpose beyond just exchanging missiles during the initial skirmishing phase. Using light troops as a screen could obscure the deployment of the heavy infantry. The importance of a general deploying his army in a manner to counter that of the enemy is highlighted by Onasander:

> 'Next the general must consider which troops to oppose to which of the enemy, and in what manner; just as a good doctor who foresees an illness of the body he must bring forward his defences and arrange his forces

> as it seems to him most advantageous; for generals are often compelled to equip and marshal their own armies with reference to the armament, nationality, and customs of the enemy.'[36]

If the light troops acted as a screening force, it made it very difficult for the enemy to deploy their army accordingly. The Battle of Ilipa (206 BC) serves as a perfect example, with Scipio using his light troops to obscure the deployment of his heavy infantry from the Carthaginians. Scipio presented his army to the Carthaginians with his Roman infantry in the centre and allied Iberians on the flanks on the days prior to the battle. On the day of the battle, however, Scipio redeployed his Roman troops on the wings and the Iberians in the centre. The new deployment was obscured by his light troops. The Carthaginians were completely caught off guard when they saw the positions of the Romans, realizing their own deployment of Libyans in the centre and their own Iberian allies on the flanks put them at a great disadvantage. Polybius states that the screening tactic 'greatly contributed to secure the victory for his [Scipio's] own men, and succeeded in putting the enemy at a considerable disadvantage'.[37]

While the role of light infantry was less decisive in pitched battles, their use as a shield, protecting the heavy infantry from the enemy's light troops was nevertheless vital. Forming a screen in front of the heavy infantry also served as a strategic tactic to obscure troop deployment.

Cost Effective

Light troops also held a certain economic advantage. The arms and armour of the heavy infantry were expensive. Van Wees and Rawlings, for instance, highlight that the Classical period hoplite's panoply cost somewhere between seventy-five and one hundred drachmas, approximately six weeks of pay for a hoplite.[38] In comparison, an archer's bow and arrows cost only two or three weeks' pay, about half that of a hoplite's panoply.[39] A javelin cost far less, at three drachmas.[40] Slings and lead slingshot were even cheaper, while stones and clay came without a price tag. The low cost of equipment allowed for greater numbers of poorer citizens to take part in warfare. Not only was the equipment of light troops less expensive, but they were paid far less than their heavy infantry counterparts. During the Peloponnesian War, skilled Thracian *peltasts* were paid one drachma (six obols) a day, compared to the hoplite's one-and-a-half drachmas (nine obols). This meant that the state could field more soldiers for far less. Furthermore, mercenaries of the Graco-Roman period were more often than not light infantry. The heavy infantry were generally citizen soldiers, which meant that they were not full-time soldiers and could not stay in the field or campaign indefinitely, having to return home to tend to their farms or other business. Mercenaries did not suffer this limitation and could remain in the field for as long as the state could pay them. This meant that during campaigns, generals could maintain a military presence and continue a war beyond the traditional campaigning season. Other light troops, especially among the large empires of the Hellenistic era, came from levied troops of subjugated people. This meant that the monarchs of the period could swell the ranks of their armies at little extra cost.

Because of this economic advantage, Graeco-Roman armies had a larger pool of manpower to draw from, could equip armies more cheaply and field them for longer periods. Furthermore, the loss of a mercenary, a poor citizen or a lowly levied soldier had a far smaller impact on the state's economy and well-being than the death of a tax-contributing citizen.

Final Remarks

Ancient and modern scholars have shown a lack of interest in light infantry of the ancient world, with the importance of their role in warfare often either overlooked, ignored or downplayed. There was a certain contempt towards the light infantry of the period. As discussed, this had very little to do with the effectiveness of the light troops in combat and the role they played in warfare, but more to do with these soldiers being generally recruited from the poorest of society, subjugated peoples, foreigners and mercenaries. Furthermore, unlike the heavy infantry, who demonstrated their *arête* or *veritas* by holding their ground against the enemy in hand-to-hand combat, light troops fought from afar and retreated when the enemy approached. This was seen as a cowardly and unmanly fighting style. Their social-economic status and fighting style made them second-class citizens, disregarded and distrusted foreigners. This is possibly best illustrated in Aristotle's claim that only those who possessed 'heavy armour' deserved political rights.[41] While they are treated with some anonymity in the sources, it does not take much reading between the lines to realize the importance of light troops in Graeco-Roman warfare. These units were invaluable to ancient armies. They were by far the most tactically diverse units of the ancient world and were assigned the greatest variety of tactical deployment. Because of this, the heaviest burdens of military campaigns were carried by the light troops. They were vital in safeguarding the army on the move, essential for any operations in difficult terrain, called upon for reconnaissance, excelled at ambushes and surprise attacks, were crucial for siege warfare and a pivotal shield for the heavy infantry during pitched battles. They were truly the forgotten and unsung heroes of ancient warfare.

Notes

Introduction

1. Thuc. 3.98.
2. Thuc. 3.98.4.
3. Plut., *Pel.*, 2.1.1.

Chapter 1

1. For the Macedonian pike-bearing line infantry *caetrati*, see Livy 44.41.8–9 and Fischer-Bovet (2014), 137–38. For an in-depth analysis on pike infantry, see Du Plessis (2022), 3–55. For Seleucid *caetrati* and their armament, see Livy 57.40.14, Du Plessis (2002), 176–78. For Achaean *caetrati*, see Livy 35.27.5 and Williams (2004).
2. For Spanish *caetrati*, see Livy 21.21.12; 23.26.1., Daly (2002), 96, 106–12, Caes., *Civ.*, 1.39.
3. Asclep. I.2; Ael. II. 7–9; Arr., *Tac.*, 3.2.
4. Asclep. I. 2.
5. D.H. 20.1–2.
6. Hdt. 9.29–30.
7. Delbrück (1990), 125, 259, 265.
8. Thuc. 2.31.2, 3.1.2; Xen., *Hell.*, 1.1.34, 2.4.32; Onas. 19; Asclep. I. 2; Ael. II.7–9; Plut., *Arat.*, 37.1; Arr., *Tac.*, 3.2; Diod. 19.40.3; Polyb. 3.104.4, 4.13.4, 5.7.11.
9. Leo IV the Wise's *Tactica* and Nikephoros Phokas' *Praecepta Militaria*, for instance, make use of the term *psiloi* to describe light troops.
10. Thuc. 4.32.3–4.
11. Onas. 17.
12. Onas. 20.2.
13. Onas. 10.8.
14. The Roman *psiloi* in Polyb. 3.104 are referred to as *euzōnoi* in 3.105, for instance.
15. For the plain of Cannae, see Polyb. 3.110.2. For the open ground needed by a phalanx, see Polyb. 18.31.5. Other examples of Polybius using *psiloi* to refer to geographical landscapes are 3.55.9 and 3.71.1.
16. Daly (2002), 108–09.
17. Liddell and Scott (1961).
18. Hom., *Il.*, 1.429.
19. Hdt, 1.72.3; Thuc. 2.97.1.
20. Xen., *Anab.*, 4.5.24, 5.4.23, 7.3.46.
21. Xen., *Anab.*, 4.5.24; 7.3.46.
22. Xen., *Anab.*, 6.3.18.
23. Liddell and Scott (1961).
24. Sekunda (2007), 343.
25. Polyb. 2.67.1–8.

26. Polyb. 11.13.1.
27. Polyb. 11.13.3–8.
28. Polyb. 5.79.6, 5.82.11.
29. Polyb. 5.79.3.
30. Polyb. 30.25.4.
31. Polyb. 3.73.1.
32. Anders (2011), 72.
33. Polyb. 1.40.7.
34. Polyb. 11.13–14.
35. App., *Iber.*, 17.
36. App., *Mith.*, 3.16.
37. Diod. 19.94.
38. Hdt. 9.63.
39. Eur., *Phoen.*, 1104.
40. Xen., *Anab.*, 4.1.28.
41. Xen., *Cyr.*, 7.5.5.
42. Xen., *Hel.*, 4.2.15.
43. Joseph., *A.J.*, 14.456; Joseph., *Bell. Jud.*, 2.512, 3.114.
44. App., *Bellum Civile*, 3.3.25.
45. WT, 738.
46. Livy 31.42.1–5.
47. Livy 31.43.1–3.
48. Livy 32.5.11.
49. Livy 33.8.3–4.
50. Livy 35.39.1–3.
51. Livy 37.41.10–12.
52. Livy 37.41.9.
53. Livy 37.18.4.
54. Livy 42.64.5.
55. Livy 44.2.9–11.
56. Livy 22.12.8–1.0
57. Livy 35.27.16.
58. Livy 38.21–22.
59. Livy 38.21.14.
60. Livy 38.21.7–8.
61. Livy 30.33.3.
62. Caes., *Gal.*, 2.10.
63. Caes., *Gal.*, 8.19.2.
64. Caes., *Gal.*, 2.19, 2.24.
65. Caes., *Gal.*, 7.65.4.
66. Caes., *Gal.*, 7.80.3.
67. Caes., *Gal.*, 8.17.2.
68. Caes., *Civ.*, 1.48.7.
69. Caes., *Civ.*, 1.83.5.
70. Caes., *Civ.*, 3.45.2–4.
71. For other passages referring to archers and light troops, see Caes., *Gal.*, 7.80.3, and *Civ.*, 3.62.2.
72. Cic., *Brut.*, 1.37.

73. Cic., *Div.*, 2.26.
74. Front., *Strat.*, 2.5.31.
75. Suet., *Jul.*, 2.66; Tac., *Ann.*, 2.8, 2.16, 13.41, 13.44.
76. Tac., *Ann.*, 2.8, 2.16, 14.33.
77. Amm. 24.6.9.

Chapter 2

1. Pain (2001).
2. Lahr et al. (2016).
3. Wilkinson (1999).
4. Farrokh, Khorasani and Dwyer (2018).
5. Latham and Paterson (1970).
6. Faris and Elmer (2010).
7. Loades (2016).
8. Hom., *Il.*, 11.385.
9. Hom., *Il.*, 11.390.
10. Hom., *Il.*, 14.479.
11. Soph., *Aj.*, 1120–1124.
12. Eur., *Her.*, 158.
13. Kelly (1979), 216–18; Pritchard (2018), 94–95.
14. Hdt. 9.72.
15. Thuc. 4.40.
16. Thuc. 6.96.
17. Hom., *Il.*, 11.395–400.
18. Hdt. 9.22.1–3.
19. Hdt. 9.60.3.
20. Thuc. 2.13.8; Aristot., *Const. Ath.*, 24.3.
21. Pritchard (2021), 410; Pritchard (2018), 96–99; Tadlock (2012), 22; Aristot., *Const. Ath.*, 24.1–2.
22. Thuc. 3.17.4; 5.47.6.
23. Pritchard (2018), 96; Pritchard (2021), 409.
24. Pritchard (2018), 90–96.
25. IG13 138. For the inscription, see Jameson (1980).
26. Thuc. 3.98.1.
27. Thuc. 3.98.1, 3.107.1, 4.129.2, 6.43.
28. Thuc. 4.55.1–2.
29. Xen., *Hell.*, 4.2.16.
30. Thuc. 3.2.2.
31. Thuc. 6.43.
32. Paus. 1.23.4.
33. Xen., *Anab.*, 1.2.9.
34. Xen., *Anab.*, 3.4.2.
35. Xen., *Anab.*, 3.4.26.
36. Xen., *Anab.*, 3.4.30.
37. Xen., *Anab.*, 3.3.7.
38. Xen., *Anab.*, 4.1.18, 4.2.28; Diod. 14.27.3.
39. Xen., *Hell.*, 3.4.16.
40. Xen., *Hell.*, 4.2.1–5.

41. Xen., *Hell.*, 4.2.14.
42. Xen., *Hell.*, 4.2.16, 4.7.6.
43. Sekunda (2010), 451, 458–59; Heckel (1992), 306–13.
44. Arr., *Anab.*, 3.12.2–3.
45. Sekunda (2010), 451.
46. Arr., *Anab.*, 2.9.3; Diod. 17.57.4.
47. Diod. 19.14.5.
48. Plut., *Pyrrhus*, 15.1.
49. Livy 37.40.8.
50. Hdt. 7.158.4.
51. Livy 22.37.9.
52. Eur., *Phoen.*, 1000–1004.
53. Thuc. 4.126.5.
54. Thuc. 1.105–106.
55. Hom., *Il.*, 13.710–715.
56. Polyb. 6.38.
57. IG i3 138.3–4.
58. Isocrates 8.46.
59. Grozer's step-by-step video of traditional bow making: https://www.youtube.com/watch?v=b_sb2Me3hdo.
60. Quoted in Latham and Paterson (1970), 94.
61. Yaden (1963); Collon (1983).
62. Hom., *Il.*, 4.105; Hom., *Od.*, 21.11, 395. For more on the composite bows in Homeric epics, see Sutherland (2001),111–20, and Thomas (1987), 44–48.
63. For 'Scythians' as a symbolic representation, see Ivantchik (2006), 197–272, and Davis (2013), 168–90.
64. Aesch., *Lib.*, 116.
65. Loades (2016), 10.
66. Karpowicz and Selby (2010), 94–132.
67. Staatliche Museen zu Berlin, # F 2588.
68. Karpowicz and Selby (2010).
69. White (2019), 21, 27–29; Zutterman (2003), 140.
70. Zutterman (2003), 140·
71. Hdt. 7.61.
72. White (2019), 30; Zutterman (2003), 140.
73. Hdt. 1.73.3.
74. Loades (2016), 72, 75.
75. Quoted in Elliot (2001), 180.
76. The drachm depicting the image of King Arsaces I (247–211 BC), for example, has a seated archer on the reverse of the coin holding what appears to be the earlier Scythian bow, with extreme angular curvatures.
77. Plut., *Crass.*, 25.6.
78. Hom., *Od.*, 12.83–84, 12.101–102, Hdt. 4.139.1, Thuc. 7.30.2.
79. McLeod (1965), 6.
80. McLeod (1965), 6.
81. Strab. 14.1.23.
82. Veg., *Mil.*, 2.23.

83. Cohen (2010), 120, 121. The Alexander Sarcophagus shows Persian warriors drawing their bows with the Mediterranean draw. It is, however, sculpted by Greeks and suggests that the Mediterranean draw was known to the Greeks.
84. Staatliche Museen zu Berlin, # F 2588.
85. Royal Museums of Art & History, # A.718.
86. Basel, Antikenmuseum und Sammlung Ludwig # BS 453.
87. University of Pennsylvania Museum, # 31–19–2.
88. Davis (2013), 11, n.31.
89. Again, it must be noted that there is plenty of iconography of medieval archers shooting their bows with the Mediterranean draw and the arrow resting on the right of the bow, and images of the arrow placed on the right does not necessary indicate the use of a thumb draw.
90. Athens National Museum, # 16384; Metropolitan Museum of Art, # 2010.147.
91. Hom., *Il.*, 4.116.
92. Hydria depicting Heracles seizing Nereus serves as an example, British Museum, # 1843,1103.47.
93. Celsus, *Med.*, 7.5.4.
94. Plut., *Crass.*, 25.5.
95. Koutserimpas, Alpantaki and Samonis (2017), 683, Table 1; Davis (2013), 125. Davis mentions six, but Koutseimpas, Alpantaki and Samonis highlight an additional three.
96. Hom., *Il.*, 11.514.
97. Arist., *Mir.*, 141.
98. Hom., *Od.*, 1.261–2.
99. Dioscorides 1.114; Hom., *Il.*, 583–585.
100. Aldrete *et al.* (2013a), 57–68·
101. Aldrete *et al.* (2013a); Aldrete *et al.* (2013b).
102. Antikenmuseum und Sammlung Ludwig # BC 480; Metropolitan Museum of Art, # 06.1021.117.
103. British Museum, # 1848,1020.45.
104. Princeton University Art Museum, # 2015–11.
105. Verwijmeren (2016), 14.
106. British Museum, # 1848,0619.8.
107. British Museum, # 1961,0710.1.
108. Laign (1965): IG i3 1032.
109. Antikenmuseum und Sammlung Ludwig # BS 459.
110. Museum of Fine Arts, Boston, # 01 .8074.
111. British Museum, # 1848,0619.8.
112. Xen., *Anab.*, 5.2.29; Polyb. 10.29.6.
113. It should be noted that several scholars have argued that even the English longbowmen focused on shooting at a much closer range than what is commonly thought. Loades (2013), 67–68, and the British Royal Armouries website (https://royalarmouries.org/stories/our-collection/longbows-of-agincourt/) serve as an example.
114. Xen., *Hunt.*, 1.18, Polyb. 31.29.6–8.
115. Onas. 17.1.
116. Hdt. 9.22.1–2.
117. Arr., *Anab.*, 1.14.1, 3.13.5.
118. Blyth, 'The Effectiveness of Greek Armour Against Arrows in The Persian War (490–479 BC)' (1977); Davis (2013), 102, 125–37.

119. Hom., *Il.*, 580–585.
120. Thomas (1987), 200–01.
121. Xen., *Hell.*, 2.4.33.
122. British Museum, # 1848,1020.45.
123. Plut., *Crass.*, 25.1.
124. Koutserimpas, Alpantaki and Samonis (2017), 683, Table 1.
125. Davis (2013), 101. The Persian Immortals (armed with bows) alone numbered 10,000. At Carrhae, the Parthians fielded 10,000 horse archers.
126. Thuc. 4.43.
127. Livy 22.37.8, 37.41.10–12, 38.21.1–4.
128. For Hiero, see Livy 22.37.8. For Eumenes, see Livy 37.39.10.
129. Thuc. 2.23.1–2.
130. *Inscriptiones Graecae* i[3] 60.9–18.
131. Thuc. 7.62.2.
132. Diod. 13.16.3.
133. Paus. 10.21.4.
134. Virt. 1.5.4.
135. The city of Iasus in Caria had towers stationed every 30 metres along its mainland wall.
136. Thuc. 3.2. Davis (2013), 110, n.382, highlights that Miletus was very active in establishing colonies in Pontus. Due to sharing borders with the Scythians, they likely adopted the bow; these archers requested by Miletus might have been colonists rather than Scythians.
137. Arr., *Anab.*, 4.2.3–4.
138. Polyb. 8.6.1; Livy 24.34.4.
139. Diod. 13.54.7.
140. Hdt. 8.52.
141. Polyb. 4.56.3.
142. Thuc. 6.96.2.
143. This led to the popular saying that each English archer carried the lives of twenty-four Scots under his belt.
144. McGeer (2008), 15: *Praecepta Militaria* 1.30.4.
145. Livy 26.4.4.
146. Thuc. 3.98.
147. Hom., *Il.*, 8.81–85.
148. Hdt. 9.22.1.
149. Hdt. 9.60.2–3.
150. Ceos. 6.2.
151. Hom., *Il.*, 8.266–272.
152. Thuc. 4.9–13.
153. Livy 31.35.3, 37.40.8.

Chapter 3

1. Rihill (2007), 2.
2. Korfmann (1973), 38.
3. Borovsky, Lankovsky, Kalichman and Belkin (2017), 10–15.
4. Veg., *Mil.*, 1.16.
5. Livy 1.43.7.
6. 1 *Sam.* 17.

7. Spence, Kelly, Londey and Phang (2016),: 183.
8. Onos. 19.3.
9. Hom., *Il.*, 13.715.
10. Hom., *Il.*, 12.155–160.
11. Hdt. 7.158.
12. Thuc. 2.81.8.
13. Thuc. 6.22.
14. Xen., *Anab.*, 3.3.16.
15. Thuc. 7.31.5.
16. Thuc. 4.100.1.
17. Xen., *Anab.*, 3.3.7.
18. Diod. 14.27.5.
19. Xen., *Anab.*, 3.4.16.
20. Xen., *Anab.*, 3.4.15.
21. Xen., *Hell.*, 4.6.7.
22. Xen., *Hell.*, 7.1.19.
23. Diod. 15.85.4.
24. Diod. 15.85.5.
25. Diod. 17.42.1, 17.59.1, 17.110.2; Arr., *Anab.*, 4.2.4–5, 4.4.3–6, 4.30.1.
26. Diod. 18.72.6.
27. Diod. 19.14.5.
28. Diod. 19.21.3.
29. Diod. 19.82.3.
30. Diod. 19.69.1.
31. Polyb. 5.52.5.
32. Potts (2014), 111; Sherwin-White and Kuhrt (1993), 55; Bar-Kochva (1989), 103.
33. Livy 37.40.8.
34. Polyb. 5.79.6.
35. Polyb. 10.30.7.
36. Livy. 35.27.5.
37. Livy 38.293.9.
38. Livy 38.29.8.
39. Pritchett (1985).
40. Polyb. 4.61.2.
41. Diod. 14.50.4.
42. Diod. 20.11.1, 20.38.3; 20.54.4.
43. Livy 22.37.9.
44. Livy 22.37.
45. Livy 27.38.12.
46. Livy 38.46.4.
47. Livy 42.58.13.
48. Sall., *Iug.*, 46.7, 57.4.
49. Caes., *Gal.*, 3.4.
50. Dio Cass. 41.59.1, 41.60.1; 43.8.2.
51. Dio Cass. 49.20.2.
52. Dio Cass. 49.26.2.
53. Diod. 5.17.1.
54. Veg., *Mil.*, 3.14.

55. For the slings found in King Tutankhamun's Tomb, see Carter (2014), 38, Pl. LXXII, A; For the sling discovered in the Tibetan fortress, see Echols (1950), 228.
56. Hom., *Il.*, 13.715.
57. Xen., *Anab.*, 3.3.18; Livy 38.29.6.
58. 1 *Sam.* 17.40.
59. Korfmann (1973), 38.
60. Xen., *Anab.*, 3.3.17.
61. Diod. 19.109.2.
62. Veg., *Mil.*, 1.16.
63. For clay slingshot dating from the eighth millennia BC found at Hassuna in Iraq and other places, see Korfmann (1973).
64. Tac., *Ann.*, 4.72–74.
65. Kelly (2012), 282.
66. Xen., *Anab.*, 3.3.16, 3.4.16.
67. Dio Cass. 49.26.2.
68. Onos. 19.3.
69. Kelly (2012), 290.
70. Kelly (2012), 292; Echols (1950), 228.
71. Polyb. 27.11.
72. Livy 42.65.7–12.
73. Bernard (1874).
74. Hollenback (2005), 79.
75. Hom., *Il.*, 11.265.
76. Hom., *Il.*, 20.285.
77. Hom., *Il.*, 16.579.
78. Hom., *Il.*, 21.404.
79. Pl., *Leg.*, 8.834.
80. Thuc. 1.106.1–2.
81. Thuc. 4.32.4.
82. Thuc. 6.69.2.
83. Xen., *Hell.*, 2.4.33.
84. Diod. 17.68.2.
85. Joseph., *Bell. Jud.*, 3.7.18.
86. Plut., *Pyrrhus,* 34.2–3.
87. Plut., *Arist.*, 19.1.
88. Veg., *Mil.*, 1.16.
89. Dio Cass. 49.26.2.
90. Finney (2005), 174–91; Vega and Craig (2009), 1266–68, Tb. 2; Skov (2013), 71–76.
91. www.slinging.org (2018).
92. Skov (2013), 48, Tb. 2.
93. Dio Cass. 49.20.2.
94. Vegetius 1.16.
95. Du Plessis (2020), 211–13.
96. Du Plessis (2020), 210–11.
97. Onas. 19.3.
98. Celsus., *Med.*, 7.5.4.
99. York and York (2011), 23.
100. Veg., *Mil.*, 1.16.

101. Hom., *Il.*, 12.154–160.
102. Hom., *Il.*, 12.285.
103. Thuc. 2.81.8.
104. Thuc. 4.93.3, 4.100.1.
105. Diod. 13.54.7, 13.56.7.
106. Diod. 17.42.1, 17.42.7.
107. Diod. 18.70.7.
108. Livy 1.41.12.
109. Livy 24.34.5.
110. App., *Mith.*, 33.
111. App., *Mith.*, 31.
112. Thuc. 7.70.5.
113. Diod. 14.50.4.
114. Diod. 20.6.3.
115. Diod. 17.42.1.
116. Livy 24.34.
117. Livy 35.27.8.
118. Flor. 1.42.1.
119. Livy 28.37.12.
120. Veg., *Mil.*, 1.16.

Chapter 4

1. The *Praecepta Militaria* (II.24) of the Byzantine Emperor Nikephoros Phokas mention that the ancient Rus were expert *akontistai*.
2. Hom., *Il.*, 16.328; Hom., *Od.*, 18.262.
3. Hdt. 8.90.2.
4. Thuc. 3.97–98.
5. Thuc. 3.104.4.
6. Thuc. 3.112.6.
7. Thuc. 4.100.1.
8. Thuc. 6.20.4.
9. Thuc. 6.67.2.
10. Thuc. 7.5.3.
11. Thuc. 7.6–7.
12. Thuc. 7.40.5.
13. Thuc. 7.78.2–6.
14. Thuc. 7.81.4.
15. Thuc. 7.81.5.
16. Thuc. 7.82.1.
17. Thuc. 7.84.1.
18. Thuc. 7.84.3–4.
19. Xen., *Anab.*, 3.3.7.
20. Xen., *Anab.*, 4.3.28.
21. Xen., *Cyr.*, 2.1.7, 2.1.16, 3.3.60, 4.3.5, 8.5.12.
22. Xen., *Hel.*, 2.4.12.
23. Xen., *Hel.*, 6.1.9.
24. Xen., *Hel.*, 3.4.16.
25. Xen., *Cyr.*, 7.3.26.

26. Polyb. 2.30.1–7.
27. Polyb. 3.65.3, 3.96.
28. Polyb. 3.72.2, 3.73.3. It is interesting to note that, as far as I am aware, the term *pezakontistai* is used only by Polybius.
29. Polyb. 5.79.11; For the Cardacian and Lydian *akontistai*, see Du Plessis (2020), 137–38.
30. Polyb. 10.30.9.
31. Onas. 17.
32. Onas. 17.
33. Arr., *Extax.*, 5; Josep., *Bell. Jud.*, 3.168, 5.297.
34. Alexander Mosaic: Museo Archeologico, Naples # 10020.
35. Polyb. 6.22.4.
36. Livy 26.4.4.
37. Xen., *Eq.*, 12.12; Theoph., *Hist. Pl.*, 3.12.2.
38. For *soliferreum*, see Silva (2013), Ch. 3, 'Lusitanian Warfare and Weapons: Spears'. A ritually 'killed' (bent out of shape as a votive offering) *soliferreum* archaeological find can be found in Museo Arqueologico Nacional, Madrid # 1940/27/AA/3.
39. Nola Tomb Frieze (Museo Archeologico, Naples # 9363).
40. Polyb. 6.22.3.
41. Asclep. I. 2.
42. British Museum, London # 1929.11–11.1.
43. Staatliche Museen zu Berlin, # 2639.
44. Asclep. 1.2; Ael. 2.7–9; Arr., *Tac.*, 3.4.
45. Asclep. 1.2.
46. Best (1969), 97.
47. Thuc. 2.29.5.
48. Thuc. 2.79.3–4.
49. Thuc. 4.28.4, 4.32.2.
50. Thuc. 4.93.3.
51. Thuc. 4.110–112.
52. Thuc. 4.123.4.
53. Thuc. 4.129.2.
54. Thuc. 5.6.4–5.
55. Thuc. 7.27.1.
56. Thuc. 7.30.2.
57. Xen., *Hel.*, 2.24.33.
58. Xen., *Anab.*, 1.2.9, 1.7.10, 1.8.5, 1.10.7, 6.2.16.
59. Xen., *Anab.*, 5.2.12.
60. Rose (2014), 136–38; Campbell and Tritle (2012), 150.
61. The Greek mercenary in the Altıkulaç Sarcophagus.
62. F498 Rose.
63. Xen., *Anab.*, 5.2.29–32.
64. Asclep. 5.1; Ael. 7.14.
65. Polyb. 2.28.7–2.30.4.
66. A Thracian *peltast* on an Athenian kylix is wielding a spear in an under-arm thrusting method. Harvard University, Arthur M. Sackler Museum, Cambridge: # 1959.219.
67. Xen., *Anab.*, 5.2.12.
68. Gordon (1958), 24
69. Xen., *Eq.*, 12.11.

70. The Alexander Mosaic and the Alexander Sarcophagus show Alexander and his elite cavalry armed with the *xiphos* sword. Alexander Mosaic: Naples, Museo Nazionale, # 10020; Cohen (2010), 120, 121.
71. Nepos, *Iphicrates*, 1.1.
72. Plut., *Pel.*, 2.1.1.
73. Nepos, *Iphicrates*, 1.3–4·
74. *Aspis*: Classical Greek hoplite shield.
75. Diod. 15.44.2–4.
76. Bugh (2006), 269–70; Sekunda (2007), 328–29; Matthew (2016), 11–19.
77. Hunt (2006), 113.
78. Bardunias and Ray (2016), 15.
79. Josh Bowmar, an all-American college javelin champion who is also a passionate spear hunter, says that the hunting range with a spear is 10 yards (Bucktin, 2019). Mitch Yonkman, another avid spear hunter, gives the prime range as 5–15 yards (Schlake, 2019).
80. Du Plessis (2002), 178–94; Matthew (2012), Ch. 9, 'The Penetrative Power of the Hoplite Spear: The Thickness of Hoplite Armour'. For the thickness of bronze body armour, see Blyth (1977), 71, Table 3.1.
81. Xen., *Hel.*, 4.5.13–17.
82. Matthews (2015), 11.

Chapter 5

1. Polyb. 5.79.4. For the Seleucid *argyraspides* as *peltasts*, see Bar-Kochva (1976),: 62–63; Du Plessis (2002), 48–55.
2. Polyb. 5.65.1–5, 5.82.3.
3. Polyb. 18.25.8–10, 18.25.1–3.
4. Livy (44.40.6) refers to the *caetrati* at Pydna as armed with the same *sarissa* as the phalanx. The *caetrati* as mentioned by Livy (31.36.1) is the Latin term for *peltasts*: '*caetratos, quos peltastas uocant*'. The *caetra* shield is the Latin term for the *peltē* shield used by Livy (28.5.11): '*pelta caetrae haud dissimils est*'.
5. Polyb 4.64.6–7; Du Plessis (2019), 167–83; (2002), 55–71.
6. Polyb. 4.75.2.
7. Polyb. 4.80.6–9.
8. Polyb. 10.49.1–5.
9. Polyb. 5.4.8–13, 10.31.13.
10. Polyb. 5.13.3–7.
11. Polyb. 5.23.8–10.
12. Polyb. 5.23.10.
13. Matthew (2016), 67–68; Campbell (2016), 48–52; Du Plessis (2019), 171–72.
14. Theoph., *Hist. Pl.*, 3.12.1–2; Asclep. 5.1; Du Plessis (2020), 5–8.
15. Polyb. 18.29.2.
16. Polyaen., *Strat.*, 2.29.2; Ael. 14.
17. Du Plessis (2020), 5–8.
18. Polyb. 18.18.3.
19. Asclep. 5.1.
20. Adam-Veleni (1993), 19 (65.6cm); Liampi (1990), 157; Hammond (1996), 365. Hammond takes an interesting approach of establishing the 2cm *daktylos* through the decorative techniques on the shield. It must be remembered that the shield find is very damaged and exact measurements are not possible.

21. Peltz (2001, 333.
22. Moormann (2000), 187–88·
23. Peltz (2001), 334·
24. Peltz (2001), 334.
25. Peltz (2001), 336; Kenneth and Wight (2010), 48. It must be noted that the Pontus shield is much larger than the *peltē*; its reference here simply serves to demonstrate connecting the bronze facing to a wooden core.
26. *Meletemata* 22, Epig. SEG 40. 524, http://epigraphy.packhum.org. On the armament of the citizen phalanx in detail, see Du Plessis (2002), 3–48.
27. Plut., *Aem.*, 18.3.
28. Penelope (2007'; Pompei, House # I 10.4; Sekunda (2012), 43–44.
29. Sekunda (2012), 43–44.
30. Sekunda (2007), 341; Bugh (2006), 270.
31. For the Stele of Zoilos, see Hatzopoulos (2001), 71, plate VI.
32. Xen., *Hel.*, 4.34.3.
33. Head (2012), 35.
34. Polyb. 5.79.4.
35. Polyb. 5.85.10; Livy 37.40.7.
36. Livy. 42.51.4–5.
37. Polyb. 10.49.2.
38. Polyb. 5.23.1–4.
39. Polyb. 8.14.8–7.
40. Polyb. 8.14.9.
41. Polyb. 8.14.9.
42. Polyb. 5.4.8–10.
43. Polyb. 10.31.13.
44. Polyb. 10.31.8–11.
45. Livy 34.19.11–13.
46. Sekunda (2012), 34.
47. Livy 37.40.7.
48. Bugh (2006), 270; Everson (2005),195–96; Head (1982), 114–15; Serrati (2013), 186–87; Sekunda (2007), 399–43. Esposito (2019), 65. The vast majority of mercenaries among the Ptolemies are depicted as *thureophoroi* – Sekunda (1994), 19–33, Colour Photos 1–3, Fig. 65–72.
49. Griffith (1935), 257.
50. Sage (1996), 178; Chaniotis (2005), 80; Trundle (2004), 105; Yalichev (1997), 199; Bar-Kochva (1976), 20, 36, 91.
51. Polyb. 5.53.8; 10.29.6.
52. Plut., *Phil.*, 9.1.
53. Du Plessis (2002), 127–33.
54. Polyb. 5.23.1–10.
55. Polyb. 11.13.1.
56. Numerous funeral stelae of mercenaries in Ptolemaic or Seleucid employ depict their deceased owners carrying *thureoi* shields; Sekunda (1994), 19–33, C.Fs. 1–3, Fig. 65–72. The arsenal at Aï Khanoum also produced a *thureos* cover, attesting its use among the Bactrians; Nikonorov (1997), 6, 10, fig. 8, 24a–b. The terracotta of two *thureophoroi*, now in the Staatliche Museen zu Berlin Preußischer Kulturbesitz, Antikensammlung #TC7696, is believed to have been found in Pergamum. The Tombstone of Eubolos

(Tagara Museum; *Lexicon of Greek Personal Names* (LGPN) 2007, III B *Eubolos* 8), a member of the Boeotian League, displays two *thureoi* over the deceased's name. Several small terracotta *thureoi* from the Tomb of Eretos, a mercenary employed by the Antigonids, show the type of shield was well known and important to these troops; Museum of Fine Arts, Boston (2017), #97.333.

57. Sekunda (2007), 341; Bugh (2006), 270.
58. Plut., *Phil.*, 9.1.
59. Bugh (2006), 270; Sekunda (2007), 399; Nefedkin (2009), 366.
60. Plutarch states that the Achaeans used *thureoi* before Philopoemen's reforms (*Phil.*, 9.1). For the Seleucid armies, see Polyb. 5.53.8 (*thureophoroi* among Molon's rebels) and Polyb. 10.29.6 (*thureophoroi* in the army of Antiochus at Mt Labus). The Tombstone of Eubolos (Tagara Museum; *Lexicon of Greek Personal Names* (LGPN) 2007: III B *Eubolos* 8), a member of the Boeotian League, displays two *thureoi* over the deceased's name. Several small terracotta *thureoi* from the Tomb of Eretos, a mercenary employed by the Antigonids, show the type of shield was well known and important to these troops (Museum of Fine Arts, Boston, 2017: #97.333). Numerous stelae of mercenaries in Ptolemaic employ depict their deceased owners carrying *thureoi* shields (Sekunda, 1994, 19–33, C.Fs. 1–3, Fig. 65–72). The arsenal at Aï Khanoum also produced a *thureos* cover, attesting its use among the Bactrians (Nikonorov, 1997, 6, 10, fig. 8, 24a–b). The terracotta of two *thureophoroi*, now in the Staatliche Museen zu Berlin Preußischer Kulturbesitz, Antikensammlung #TC7696, is believed to have been found in Pergamum.
61. Sekunda (1994a), 19–22. While Sekunda gives a compelling argument that the figures on the stelae represent 'Romanized' Ptolemaic infantry, I am unconvinced. The shields are far too narrow and small to be the Roman *scutum*, which was essential for the Roman style of fighting. Neither do the spears depicted in the stelae correspond with the Roman *pilum*, nor are any of the soldiers wearing any greaves in the depictions, as Roman soldiers did on their leading leg. Very few of the stelae show the deceased to wear armour; according to Polybius, even the lightest of the line infantry, the *Hastati*, wore some form of body-armour. I am thus of the opinion that they simply represent Hellenistic *thureophoroi*.
62. The vast majority of the *thureoi* on the funerary stele in the Sidon Collection showing Ptolemaic mercenaries are between shoulder- and knee-length (Sekunda, 1994, 19–33, Colour Photos 1–3, Fig. 65–72), but larger versions did exist which reached from the shoulder all the way to the ankle (Sekunda, 1994b, 75, fig. 23).
63. Plut., *Phil.*, 9.1.
64. Staatliche Museen zu Berlin Preußischer Kulturbesitz, Antikensammlung #TC7696.
65. The Roman *scutum* shield, which bears major similarities to the *thureos*, had, according to Polybius (6.23.1–6), a wooden core which was covered in canvas and ox-hide.
66. Bardunias and Ray (2016), 13
67. *Ptolemaic Paintings and Mosaics and the Alexandrian Style* (Brown, 1957, 28, no. 27, pl. xx, 1).
68. Many of the *thureophoroi* depicted on the Sidon Stelae and those on an Alexandrian fresco (Hanfmann, 1992, 242–55) are clearly shown armed with swords as side arms.
69. Istanbul Archaeological Museum # 1490; Sekunda (1994a), 27–28, C.F. 3, fig. 71.
70. The Stele of Saëttas (Istanbul Archaeological Museum # 1489), as well as other terracotta figures from Egypt, carries the type of sword; Sekunda (1994a), 74, fig. 91–93. Sekunda (1994a), 26, suggests that it is a prototype of a *gladius*. The sword blade,

however, bears few similarities to the *gladius hispaniensis* in its shape, but the dramatic tapering of the blade is akin to some Greek *parazonium* daggers.

71. Sekunda (1994a), 21, 26, 27.
72. Hanfmann (1992), 242–55; Sekunda (1994a), 74, fig. 84–85. While the mural comes from the Roman period, it is believed to be a copy from the third or second century BC.
73. Istanbul Archaeological Museum # 1167.
74. Polyb. 4.12.3, 11.11.4–5, 11.14.1, 11.15.5.
75. Du Plessis (2020), 127–32.
76. Polyb. 5.53.8.
77. Polyb. 5.82.10.
78. Polyb. 11.15.1–5.
79. Polyb. 11.14.1.
80. Polyb. 5.84.9.
81. Polyb. 16.19.1.
82. App., *Syr.*, 4.17–18.
83. Polyb. 10.29.6.
84. Polyb. 6.22.7.
85. Livy 26.4.3–10.
86. Livy 26.4.4.
87. Livy 26.4.7–8.
88. Livy 26.4.10.
89. Anders (2011), 67.
90. Livy 21.55.11.
91. 'praefixa ferro quale hastis velitaribus inest'.
92. Anders (2011), 96.
93. Polyb. 6.35.5.
94. Polyb. 9.3.1–4.
95. While Polybius uses *euzōnoi* in some of these references, there is little doubt the *euzōnoi* among the Roman forces are the *velites*.
96. Livy. 27.18.5–7.
97. Livy 27.18.11–15.
98. Livy 38.21.3.
99. Livy 38.21.11–14.
100. Livy 31.35.6–7.
101. Polyb. 6.22.3.
102. Polyb. 6.39.4.
103. Sall., *Jug.*, 46.7.
104. Polyb. 6.22.
105. Strabo 4.4.3; Plut., *Sulla*, 18.
106. We cannot be certain about the measurements used by Polybios, which is why only approximate values are given.
107. Livy 26.4.4.
108. Livy 26.4.4, 38.201.
109. Lucilius 7.290; Livy 26.4.4.
110. Cic., *Brut.*, 127, 'Like the javelins of the *velites* with the *amentum*' (*Ut hastae veltibus amentatae*).
111. Polyb. 3.73.3.
112. Polyb. 6.22.1; Livy 31.15.6–7, 38.21.12–13.

113. Quesada Sanz (1997), 256·
114. Polyb. 6.23.6.
115. Livy 31.34.4.
116. Polyb. 6.22.2; Livy 31.35.6.
117. Livy 38.21.12.
118. Bishop (2020), 8–12, 32–33.
119. Pliny, *HN.*, 16.209.
120. Chapter 12 of the *Fóstbræðra Saga.*
121. Polyb. 1.40.13,
122. Livy 21.55.11.
123. Livy 21.56.1.
124. Polyb. 15.12.3–4.

Chapter 6

1. Griffith (1935), 245; Yalichev (1997), 200.
2. At the Battle of Raphia, both Ptolemy and Antiochus employed Cretans (Polyb. 5.82.3). At Magnesia, Cretans also served on both sides of the line (Livy 37.40.8, 37.41.10).
3. Livy 43.7.1–4.
4. Paus. 4.8.3, 4.10.1, 4.19.4–6.
5. Hdt. 7.171.
6. Thuc. 6.25.2.
7. Paus. 1.29.5.
8. Xen., *Anab.*, 1.2.9.
9. Xen., *Hel.*, 4.2.16.
10. Xen., *Hel.*, 4.7.6.
11. Xen., *Hel.*, 7.5.1; Diod. 15.82.6–15.83.2.
12. Diod. 17.17.4.
13. Gaebel (2002), 195.
14. Diod. 20.85.3.
15. *Bulletin de correspondence hellenique*(1889), vol. 13, p.47.
16. Polyb. 4.55.5, 4.61.2.
17. Polyb. 4.55.5.
18. Livy 33.14.3.
19. Livy 42.51.7.
20. Polyb. 5.65.7, 5.79.10.
21. Polyb. 5.53.3, 10.29.6, 10.30.9.
22. Livy 37.40.8.
23. For the various troop types represented at the parade, see Polyb. 30.25.1–12.
24. Joseph., *AJ.*, 8.86; 1 Macc. 10.67.
25. Polyb. 31.17.8.
26. Plut., *Pyrrhus*, 27.
27. Plut., *Cleom.*, 6.3, 21.3; Plut., *Arat.*, 63.
28. Polyb. 5.36.4–5.
29. Polyb. 4.80.6.
30. Livy 34.27.
31. Livy 33.14.6; Polyb. 38.13.1.
32. Livy 42.55.10.
33. Orientis Graeci Inscriptiones Selectae. i. 270.

34. Livy 37.39.10; App., *Syr.*, 31.
35. Livy 38.21.2.
36. Griffith (1935) 176.
37. Polyb. 3.75.7–8.
38. Livy 22.37.9.
39. Livy 33.3.10.
40. Livy 33.3.10.
41. Plut., *Gracch.*, 16.3.
42. Caes., *Gal.*, 2.7; App., *BC.*, 2.49.
43. Polyb. 5.36.4–5.
44. Plut., *Lyc.*, 4.
45. Strab. 10.4.16.
46. Chaniotis (2005), 12.
47. Chaniotis (2005), 81.
48. Dittenberger (1915–1924), SIG (*Sylloge Inscriptionum Graecarum*) 2.581.
49. Polyb. 5.53.3.
50. Diod. 74.4–6.
51. Pl., *Leg.*, 1.625 c–d.
52. Polyb. 6.46.2.
53. Thuc. 7.57.9.
54. Polyb. 8.6.8.
55. Griffith (1935), 169.
56. Joseph., *AJ.*, 13.135.
57. Diod. 33.4.2.
58. Polyb. 4.8.11.
59. For Lagoras' scheme and the capture of Sardis, see Polyb. 7.15–18.
60. Polyb. 8.16.7.
61. For a detailed account of the capture of Achaeus, see Polyb. 8.15–25.
62. Polyb. 5.3.1, 5.65.7, 5.79.10.
63. Walbank (1957), 540; Griffith (1935), 119.
64. Launey (1951), 257, 284.
65. Head (1982), 100; Sekunda (1994), 25; see also Sekunda in Phang *et al.* (2016), 114.
66. Spyridakis (1992), 299–307; Yalichev (1997), 200.
67. Taylor (2013), 56.
68. Xen., *Anab.*, 5.2.28–32; Plut., *Aem.*, 32.4.
69. Polyb. 10.29.6.
70. Polyb. 5.79.10; 5.82.10.
71. Kallimachos, *Epig.*, 38.
72. Xen., *Anab.*, 3.3.7.
73. Xen., *Anab.*, 3.4.17.
74. For the importance of arrow spines, see Easton Archery website, https://eastonarchery.com/2014/07/making-sense-of-arrow-spine/.
75. Xen., *Anab.*, 5.2.29.
76. Polyb. 10.29.6.
77. Plut., *Aem.*, 32.6.
78. Anders (2011), 62, fig. 5.
79. Anders (2011), 61, fig. 4.
80. Anders (2011), 61–62, fig. 4; Sekunda (2003), 77–80.

81. Asclep. 1.5.
82. Antikenmuseum und Sammlung Ludwig # BS 459; Penelope (2007'; Pompei, House # I 10.4; Sekunda (2012), 43–44.
83. Paris on the Aphaia West Pediment, the archers on the Nereid Frieze and the kneeling archer by the Pithos Painter, to name but a few.
84. Xen., *Anab.*, 3.3.7.
85. Anders (2011), 62, fig. 5.
86. Kelly (2012), 273–311.
87. Anders (2011), 61, fig. 3.
88. Polyb. 4.7.1.
89. Polyb. 4.71.11–13.
90. Pl., *Leg.*, 1.625d.
91. Xen., *Anab.*, 4.8.27.
92. Livy 31.36.8.
93. Livy 37.41.11.
94. Xen., *Hel.*, 7.5.1; Diod. 15.82.6–15.83.2.
95. Xen., *Anab.*, 4.2.28.
96. Xen., *Anab.*, 5.2.32.
97. Livy 31.39.12.
98. The *rhomphaia* was a type of polearm with a long, straight or slightly curved, slender blade, which was either the same length as the staff or slightly longer. The blade was ideal for both hacking and thrusting. Archaeological finds show the weapon to be of various lengths, with an average of more than 100cm.
99. Plut., *Cleom.*, 6.3.
100. Xen., *Hel.*, 4.7.6.
101. Plut., *Pyrrh.*, 29.4. Plutarch uses the term '*bélos*', meaning missile; various translations have recorded it as javelin, while others translated the term as arrow. Since Cretans are in the vast majority of the literature referred to as archers, and never as javelinmen, I would like to think of the missile being an arrow.
102. Livy 31.39.12.
103. Livy 37.41.9–11.
104. Diod. 5.17.1; Polyb. 3.33.11.
105. Livy 28.37.11. Livy clearly exaggerates the lack of other weapon types, but it does emphasize the importance of the weapon on the Balearic Islands.
106. Strab. 3.5.1.
107. Veg., *Mil.*, 1.16.
108. Strab. 3.5.1; Diod. 5.18.4; Florus 1.43.8; Veg., *Mil.*, 1.16.
109. Diod. 5.18.4.
110. Diod. 5.17.4.
111. Livy 28.37.9.
112. Diod. 13.80.2.
113. Diod. 19.106.2.
114. Polyb. 3.33.11, 3.33.16; Livy 21.21.2.
115. Livy 21.55.2.
116. Livy 22.4.3.
117. Livy 28.37.9.
118. Anders (2011), 49.
119. Caes., *Gal.*, 2.7; Sall., *Iug.*, 105.

120. www.tourism-mallorca.com.
121. Skov (2013), 49.
122. Diod. 5.18.3; Florus 1.43.8.
123. Strab. 3.5.1; Diod. 5.18.3.
124. Strab. 3.5.1.
125. Verg., *Georgics.*, 1.309.
126. Veg., *Mil.*, 3.14.
127. Livy 38.29.6.
128. Livy 38.29.6.
129. Livy 28.37.
130. Daly (2002), 107.
131. For the clay slingshot excavated at Carthage, see fig. 14.
132. Diod. 5.18.3.
133. Diod. 19.109.2.
134. Plant (2004), 141.
135. Diod. 5.18.3.
136. Scene LXVI.28, 31.
137. Strab. 5.18.3.
138. Strab. 3.5.1.
139. Diod. 5.18.3.
140. Diod. 5.18.3.
141. Diod. 5.18.3.
142. Diod. 19.109.1–4.
143. Diod. 5.18.3.
144. Diod. 13.54.7.
145. Caes., *Gal.*, 2.6.

Chapter 7

1. Thuc. 4.29.3–4.
2. Thuc. 4.32.1.
3. Thuc. 4.32.3.
4. Thuc. 4.32.4.
5. Thuc. 4.33.1.
6. Thuc. 4.33.2.
7. Thuc. 4.34.1–2.
8. Thuc. 4.34.2.
9. Thuc. 4.34.2.
10. Thuc. 4.34.3.
11. Thuc. 4.35.4.
12. Thuc. 4.36.3.
13. Thuc. 4.38.3.
14. Thuc. 4.38.5.
15. Thuc. 4.40.1.
16. Thuc. 5.14.3.
17. Thuc. 4.41.2.
18. Thuc. 4.41.3, 4.55.1.
19. Thuc. 4.55.3–4.
20. Thuc. 4.55.2.

21. App., *Syr.*, 7.1.
22. Polyb. 18.29.2.
23. Asclep. 5.1.
24. Polyb. 6.31.5.
25. Polyb. 10.30.2–3.
26. Polyb. 10.30.3.
27. Polyb. 10.29.5.
28. Polyb. 10.30.9.
29. Polyb. 10.29.6.
30. Polyb. 10.30.7.
31. Polyb. 10.30.9.
32. Polyb. 10.31.2.
33. Polyb. 10.31.3.
34. Polyb. 10.31.3.
35. Plut., *Flam.*, 8.4.
36. Polyb. 10.30.5.
37. Du Plessis (2020), 71–114.
38. Grainger (2010), 80–81; Heinen (1984), 423.
39. Orientis Graeci Inscriptiones Selectae # 223. For a tax for paying for hostages, see Rankin (1987), 189. For a war tax, see Walbank (1984), 194.
40. Livy 38.12.8.
41. Livy 38.19.3–5.
42. Livy 38.20.1–3.
43. Livy 38.21.4–5.
44. Livy 38.21.6–7.
45. Livy 38.21.7–12.
46. Livy 38.21.13.
47. Livy 38.22.7.
48. Livy 38.22.8.
49. Livy 38.23.1.
50. Livy 38.23.3.
51. Livy 38.23.6–8.
52. Livy 38.23.8; App., *Syr.*, 9.42.
53. Livy 38.25.16.
54. Livy 38.26.5–7.
55. App., *Syr.*, 9.42.
56. Livy 38.27.1.
57. Livy 38.46.4.
58. Livy 38.17.2.
59. Strab. 4.4.2.
60. Just. 24.4.

Chapter 8

1. Speidel (2004), 128.
2. The portrait of Captain Thomas Lee (1594) depicts him in Irish dress and wielding a javelin with a throwing strap fixed to the shaft (Tate Museum, London: # T03028).
3. In the foreground, bottom centre of the Alexander Mosaic (Naples, Museo Nazionale, # 10020), a light javelin is visible, which clearly has an *ankylē* strap fixed to the shaft.

The Nola Tomb Frieze shows a Samnite warrior holding javelins with strowing straps attached (Museo Archeologico in Naples # 9363). The javelin with which Philopoemen was struck through the thigh was described by Plutarch as thrown with an *ankylē* (Plut., *Philo.*, 6.4). The peltasts in Xenophon's army are mentioned to have made use of the *ankylē* (Xen., *Anab.*, 4.2.28, 5.2.12). Livy (37.41.3) refers to the numerous light troops at Magnesia and how these were at a disadvantage due to the damp weather. Amongst these were the *akontistai* or javelinists whose *ankylē* thongs were too wet to use effectively. I have also argued that the evidence shows that the *velites* used throwing straps for their javelins. See 5.6.1: The Grosphus Javelin.

4. For General Reffye's experiments preformed for Napoleon, see Harris (1963), 30.
5. The Panathenic amphora dating to 525 BC, the red-figure cup from the Paravey Collection in the Louvre and an Attic red figure drinking cup from Munich: # 2667 serve as but a few examples.
6. British Museum, London # B37.
7. Both Harris (1963), 30, and Murray (2012), 4–5, believe that the *ankylē* used in a sports context was merely wrapped around the shaft, rather than being fixed to it.
8. Plut., *Philo.*, 6.4–6.
9. Harris (1963), 28–29; Hunt (2006), 120.
10. Robertson (1941), 412–13, Pl CXXVII. While Robertson suggests that these are *sarissa* heads, the socket of the head measures a mere 1.5cm, meaning the shaft of the *sarissa* was tapered down to an unrealistic thinness to accommodate the head. Such a thinly tapered shaft would not stand the strenuous tests of phalanx combat, so the heads seem very unlikely to be those of a *sarissa*. The sockets were also too broad to be those ofarrowheads. It is thus far more likely that they were javelin heads. The size of those from Olympus also corresponds far better with the size of other javelin heads in iconographical evidence from various cultures.
11. Theoph., *Hist.*, Pl. 3.12.1–2; Xen., *Eq.*, 12.12.
12. Meranti has a density of 750, while that of cornel is 710. For a comparison between meranti and cornel wood (dogwood), see https://www.engineeringtoolbox.com/wood-density-d_40.html, online article 'Density of Wood Spieces' (Toolbox, 2004).
13. Polyb. 6.22.4; Livy 26.4.4.
14. Strab. 4.4.3.
15. Daly (2002), 97; Museo Archeologico, Naples # 9363.
16. 'About a finger's breadth in thickness', Polyb. 6.22.4.
17. Plut., *Philo.*, 6.4–6.
18. It is interesting to note that the javelin shafts found at Nydam, Denmark, had bronze rivets, which served the same function; Gardiner (1907), 256.
19. Since no examples or literary evidence exist for the length of the *ankylē*, I based my length on that of the throwing strap used by the New Caledonian tribesmen; Gardiner (1907), 257.
20. I obtained my provincial colours (Mpumalanga, South Africa) for the javelin.
21. For bow hunting chart on the required KE to hunt various game, see https://www.longrangehunting.com/articles/arrow-speed-and-kinetic-energy.491/; Henry (2011).
22. Matthew (2012), Ch. 9, 'The Penetrative Power of the Hoplite Spear: The Thickness of hoplite armour'; for the thickness of bronze body armour, see Blyth (1977), 71, Table 3.1.
23. See note 309 above.
24. Xen., *Hel.*, 4.5.13–17.

25. Celsus., *Med.*, 7.5.4; Onas. 19.3.
26. Onas. 19.3.
27. Onas. 19.3; Dio Cass. 49.26.2; Diod. Sic. 19.109.2.
28. Blunt Criterion (BC) serves as an energy-based model developed by Sturdivan at the Military Biophysics Lab at Aberdeen Proving Ground, Maryland, to predict injury risk and the severity of blunt force trauma. The Abbreviated Injury Scale (AIS) is an anatomical scoring system which is a reasonably accurate way of ranking the severity of injury.
29. It must be noted that term 'Greek' sling of 'triple construction' is only for the ease and purposes of the research; while this was a popular design among the Greek, it far predates Classical Greek culture. The sling find from the tomb of Tutankhamun serves as a much earlier example.
30. Korfmann (1973), 39.
31. Frank *et al.* (2011), 40.
32. Diod. 5.18.3.
33. Sturdivan, Viano and Champion (2004), 651; Bir and Viano (2004), 1218; Frank *et al.* (2011), 35.
34. Bir and Viano (2004), 1221; Borovsky *et al.* (2017), 6.
35. Pignolet *et al.* (1990), 13–21; Sturdivan *et al.* (2004), 653; Frank *et al.* (2011), 39; Lynnerup (2001), 45–51.
36. Sturdivan *et al.* (2004), 653–54; Frank *et al.* (2011), 39.
37. Wong (2011), 5, Table 1.
38. Ulman and Stalnaker (1986), 117, Table 5.
39. While the Corinthian helmet had fallen out of favour by the Hellenistic period, it still serves as good example of helmets of the time.
40. Corinthian helmets tended to be between 1mm and 2 mm thick; Jarva (2013), 402. Pompeian bronze gladiator helmets from the first century measured between 1mm and 3mm, with an average of 1.5mm; Junkelmann (2000), 40. Bronze cuirasses and greaves averaged 1mm thick; Treister (2001),115; Blyth (1977), 71, Table 3.1.
41. Livy 22.49.2–3.
42. Onos. 19.3.
43. Xen., *Anab.*, 3.3.7.
44. Xen., *Anab.*, 3.4.17.
45. Xen., *Anab.*, 3.4.15.
46. Baitinger (2001), 10–11, nn. 59–83.
47. Examples of bronze Type IA5 arrowheads appear across Greece, being excavated at Olympia, Olynthos, Delphi, Delos, Kalapodi, Lindos, Nemea and Perachora. Iron versions have been excavated at Athens, Egypt and Syria.
48. Szabó, Kiss and Tóth (2021), 46–47; Karpowicz and Selby (2010), 94–132.
49. Coulston (1985), 240.
50. Szabó, Kiss and Tóth (2021), 46–47.
51. Szabó, Kiss and Tóth (2021), 46–47.
52. Karpowicz and Selby (2010), 130.
53. Loades (2016), 72; Elliot (2001), 180.

Chapter 9

1. Lee (2007), 144.
2. Lee (2007), 167.

3. Veg., *Mil.*, 1.9.
4. Veg., *Mil.*, 3.6.
5. Xen., *Anab.*, 4.1.6.
6. Livy. 35.27.15.
7. Sall., *Jug.*, 46.100.
8. Polyb. 5.61.8; 10.49.2.
9. Polyb. 2.34.6–9.
10. Polyb. 5.68.10.
11. Polyb. 10.49.1–5.
12. Livy. 35.38. 10–11.
13. Caes., *Civ.*, 1.70.
14. Onas. 18.
15. Polyb. 4.75.2–4.
16. Polyb. 3.94.6.
17. Polyb. 3.101.5–11.
18. Caes., *Gal.*, 2.10.
19. Livy. 35.27.15–16.
20. Xen., *Hel.*, 2.4.33.
21. Xen., *Hel.*, 4.5.13–17.
22. Dio. 49.6.
23. Thuc. 3.122.6.
24. Xen., *Anab.*, 4.6.17.
25. Polyb. 5.13.3–7.
26. Polyb. 8.14.8–7.
27. Livy. 35.27.5–8.
28. Polyb. 3.110.5.
29. Polyb. 5.95.8.
30. Polyb. 11.21.7.
31. Caes., *Civ.*, 3.75.
32. Livy. 37.42.7.
33. Livy. 37.41.9.
34. Xen., *Eq.*, 5.13.
35. Xen., *Eq.*, 8.19.
36. Onas. 30.
37. Polyb. 11.22.3.
38. Van Wees and Rawlings, in Rihll (2014), 7.
39. Rihll (2014), 7.
40. Rihll (2014), 3.
41. Aristot., *Pol.*, 4. 1297.b 1–2.

Bibliography

Adkins, L., and Adkins, R., *Handbook to Life in Ancient Greece* (Oxford: Oxford Univerity Press, 1997).

Aeschylus, *Libation Bearers* (Cambridge, MA: Harvard University Press, Loeb Classical Library, 1926).

Allen, S., *Lords of Battle: The World of the Celtic Warrior* (Oxford: Osprey Publishing, 2007).

Anders, A., *Roman Light Infantry and The Art of Combat* (Cardiff: Cardiff University, 2011).

Baitinger, H., *Die Angriffswaffen aus Olympia: Die Angriffswaffen aus Olympia 29* (Berlin: De Gruyter, 2001).

Bardunias, P., and Ray, F.J., *Hoplites at War: A Comprehensive Analysis of Heavy Infantry Combat in the Greek World, 750–100 BCE* (Jefferson: McFarland & Company, 2016).

Bar-Kochva, B., *Judas Maccabaeus: The Jewish Struggle Against the Seleucids* (Cambridge: Cambridge University Press, 1989).

Bar-Kochva, B., *The Seleucid Army: Organization and Tactics in the Great Campaigns* (Cambridge: Cambridge University Press, 1976).

Bernard, A., 'Le kestre ou kestrosphendone', *Revue Archeologique*, 27 (2) (1874), 37–38.

Best, J., *Thracian Peltasts and Their Influence on Greek Warfare* (Groningen: Wolters-Noordhoff, 1969).

Bir, C., and Viano, D.C., 'Design and injury assessment criteria for blunt ballistic impacts', *Journal of Trauma – Injury, Infection and Critical Care*, 57 (6), 1218–1224.

Bishop, M.C., *Roman Shields* (Oxford: Osprey Publishing, 2020).

Blyth, P.H., *The Effectiveness of Greek Armour Against Arrows in The Persian War (490–479 BC)* (Reading: University of Reading, 1977).

Borovsky, I., Lankovsky, Z., Kalichman, L., and Belkin, V., 'The traumatic potential of a projectile shot from a sling', *Forensic Science International*, 272 (2017), 10–15.

Brown, B.R., *Ptolemaic Paintings and Mosaics and the Alexandrian Style* (Massachusetts: Cambridge University Press, 1957).

Brownson, C., *Xenophon: Anabasis*, ed. J. Dillery (Cambridge, MA: Harvard University Press, Loeb Classical Library, 1998).

Brownson, C.L., *Xenophon: Hellenica* (Cambridge, MA: Harvard University Press, Loeb Classical Library, 1918).

Brunt, P.A., *Arrian: Campaigns of Alexander* (Cambridge, MA: Harvard University Press, Loeb Classical Library, 1983).

Bucktin, C., 'Hunter who used 7ft spear to kill bear says people should be "ashamed" of attempts to stop him', retrieved 20 March 2019 from *The Mirror* website, https://www.mirror.co.uk/news/world-news/hunter-who-used-7ft-spear-8636150.

Bugh, G., 'The Theseia in Late Hellenistic Athens', *Zeitschrift Für Papyrologie Und Epigraphik*, 83 (1990), 20–37.

Bugh, G., 'Hellenistic Military Developments', in G. Bugh (ed.), *The Cambridge Companion to The Hellenistic World* (Cambridge: Cambridge University Press, 2006).

Bury, R.G., *Plato's Laws* (Cambridge, MA: Harvard University Press, Loeb Classical Library, 1968).
Campbell, B., and Tritle, L. (eds), *The Oxford Handbook of Warfare in the Classical World* (Oxford: Oxford Univerity Press, 2012).
Cary, E., *Dionysius of Halicarnassus: Roman Antiquities* (Cambridge, MA: Harvard University Press, Loeb Classical Library, 1950).
Cary, E., and Foster, H., *Cassius Dio: Roman History* (Cambridge, MA: Harvard University Press, Loeb Classical Library, 1914).
Chaniotis, A., *War in The Hellenistic World: A Social and Cultural History* (Oxford: Blackwell Publishing, 2005).
Charles, B., and Herschel, C., *Stratagems. Aqueducts of Rome* (Cambridge, MA: Harvard University Press, Loeb Classical Library, 1925).
Collon, D., 'Hunting and Shooting', *Anatolian Studies*, 33 (1983), 51–56.
Coulston, J.C., 'Roman Archery Equipment', in M.C. Bishop (ed.), *The Production and Distribution of Roman Military Equipment: Proceedings of the Second Roman Military Equipment Research Seminar* (Oxford: British Archaeological Reports Oxford Ltd, 1985).
Daly, G., *Cannae: The Experience of Battle in the Second Punic War* (London: Routledge, 2002).
Davis, T.A., 'Archery in Archaic Greece', thesis, Columbia University (2013).
DE MATERIA MEDICA of Dioscorides (Johannesburg: IBIDIS Press, 2000).
Delbrück, H., *Warfare in Antiquity, History of The Art of War* (Lincoln/London: University of Nebraska Press, 1975).
DeVoto, J.G., *Arrian: Tactical Handbook* (Chicago: Ares Publishers, 1993).
Du Plessis, J.C., '"Synaspismos" and Its Possibility in the Macedonian Styled Phalanx', *Akropolis*, 3 (2019), 167–183.
Du Plessis, J.C., *The Seleucid Army of Antiochus the Great Weapons, Armour and Tactics* (Philadelphia: Pen & Sword Military, 2022).
Echols, E., 'The Ancient Slinger', *The Classical Weekly*, 43 (15) (1950), 227–230.
Elliot, M., *The Manchu Way* (Red Wood City: Stanford University Press, 2001).
Esposito, G., *Armies of the Hellenistic States 323 BC to AD 30* (Barnsley: Pen & Sword Military, 2019).
Faris, N.A., and Elmer, R.P., *Arab Archery, An Arabic Manuscript Of About A.D. 1500: A Book On The Excellence Of The Bow And Arrow And The Description Thereof* (Whitefish: Kessinger Publishing, 2010).
Farrokh, K., Khorasani, M.M., and Dwyer, B., 'Depictions of archery in Sassanian silver plates and their relationship to warfare', *Revista de Artes Marciales Asiáticas*, 13 (2) (2018), 82–113.
Fischer-Bovet, Christelle, *Army and Society in Ptolemaic Egypt* (Cambridge: Cambridge University Press, 2014).
Flower, M., *Xenophon: Anabasis* (Oxford: Oxford Univerity Press, 2012).
Foster, E., *Florus: Epitome of Roman History* (Cambridge, MA: Harvard University Press, Loeb Classical Library, 1929).
Fowler, H., *The Works of Lucian of Samosata* (Charleston: Nabu Press, 2009).
Frank, M., Bockholdt, B., Peters, D., Lange, J., Grossjohann, R., Ekkernkamp, A., and Hinz, P., 'Blunt Criterion trauma model for head and chest injury risk assessment of cal. 380 R and cal. 22 long blank cartridge actuated gundog retrieval devices', *Forensic Science International*, 208 (1–3) (2011), 37–41.
Gardiner, N., 'Throwing the Javelin', *The Journal of Hellenic Studies*, 27 (1907), 249–273.
Godley, A.D., *Herodotus: The Persian Wars Vol II* (Cambridge, MA: Harvard University Press, Loeb Classical Library, 1921).

Grainger, J.D., *The Syrian Wars*, Vol. 320 (Boston: Brill Publishing, 2010).
Granger, F., *On Architecture* (Cambridge, MA: Harvard University Press, Loeb Classical Library).
Graves, R., *Tranquillus Gaius Suetonius. The Twelve Caesars* (Baltimore: Penguin Books, 1957).
Griffith, W., and Glenn, W., *Euripides: Helen, The Phoenician Women, Orestes*, 3rd ed. (Chicago: University of Chicago Press, 2013).
Hammond, M., *Thucydides: The Peloponnesian War* (Oxford: Oxford Univerity Press, 2009).
Harris, H.A., 'Greek Javelin Throwing', *Greece & Rome*, 10 (1) (1963), 26–36.
Head, D., *The Spartan Warrior* (Oxford: Osprey Publishing, 2012).
Heckel, W., *The Marshals of Alexander's Empire* (London: Routledge, 1992).
Heinen, H., 'The Syrian-Egyptian Wars and the new kingdoms of Asia Minor', in F. Walbank, A. Astin, M. Frederiksen and R. Ogilvie (eds), *The Cambridge Ancient History: The Hellenistic World*, 2nd edn., pp.412–445 (Cambridge: Cambridge University Press, 1984).
Henry, J., 'Arrow Speed and Kinetic Energy', retrieved 25 February 2018 from *Long Range Hunting Online Magazine* website, https://www.longrangehunting.com/articles/arrow-speed-and-kinetic-energy.491/.
Hollenback, G.M., 'A new reconstruction of the kestros or cestrosphendone', *Arms & Armour*, 2 (1) (2005), 79–86.
Holt, A. (trans.), *Theophrastus: Enquiry into Plants Vol I* (Cambridge, MA: Harvard University Press, 1968).
Hunt, P., 'Military Forces', in P. Sabin, H. Van Wees and M. Whitby (eds), *The Cambridge History of Greek and Roman Warfare – Volume I: Greece, the Hellenistic World and the Rise of Rome*, pp.108–146 (Cambridge: Cambridge University Press, 2006).
Illinois Greek Club, *Aeneas Tacitcus Asclepiodotus Onasander* (Cambridge, MA: Harvard University Press, Loeb Classical Library, 1928a).
Illinois Greek Club, *Asclepiodotus: Tactics* (Cambridge, MA: Harvard University Press, Loeb Classical Library, 1928b).
Ivantchik, A.I., '"Scythian" archers on archaic attic vases: Problems of interpretation', *Ancient Civilizations from Scythia to Siberia*, 12 (3–4) (2006), 197–272.
Jackson, J., *Tacitcus: Annals* (Cambridge, MA: Harvard University Press, Loeb Classical Library, 1937).
Jameson, M.H., 'Apollo Lykeios at Athens', *Arkhaiognosia*, 1 (1980), 213–236.
Jarva, E., 'Arms and Armour: Part 1 – Arming the Greeks for Battle', in B. Campbell and L. Tritle (eds), *The Oxford Handbook of Warfare in the Classical World*, 2013th ed., pp.1424–1504 (Oxford: Oxford Univerity Press, 2013).
Jones, W., and Ormerod, M., *Pausanias: Description of Greece* (London: Harvard University Press, Loeb Classical Library, 1918).
Junkelmann, M., 'Heroes of the Amphitheatre', in E. Kohne and C. Ewigleben (eds), *Gladiators and Caesars: The Power of Spectacle in Ancient Rome*, pp.31–74 (Berkeley: University of California Press, 2000).
Karger, B., Sudhues, H., Kneubuehl, B.P., and Brinkmann, B., 'Experimental arrow wounds : ballistics and traumatology', *Journal of Trauma: Injury, Infection, and Critical Care*, 45 (3) (1998), 495–501.
Karpowicz, A., and Selby, S., 'Scythian Bow From Xinjang', *Journal of Society of Archer-Antiquaries*, 53 (2010), 94–132.
Kaster, R., *Cicero: Brutus & Orator* (New York: Oxford Univerity Press, 2020).
Kelly, A., 'The Cretan Slinger At War – a Weighty Exchange', *Annual of the British School at Athens*, 107 (October 2012), 273–311.

Ker, R.E., *Simonides of Ceos* (Belfast: Queen's Univeristy of Belfast, 1930).
Korfmann, M., 'The Sling as a Weapon', *Scientific American*, 229 (4) (1973), 34–42.
Koutserimpas, C., Alpantaki, K., and Samonis, G., 'Trauma management in Homer's Iliad', *International Wound Journal*, 14 (2017), 682–684.
Krentz, P., and Wheeler, E., *Polyaenus: Strategems of War* (Chicago: Ares Publishers, 1994).
Laing, D.R., *A New Interpretation of the Athenian Naval Catalogue* (Cincinnati: University of Cincinnati, 1965).
Latham, J.D., & Paterson, W.F., *Saracen Archery* (London: The Holland Press, 1970, trans. of 1368 original).
Launey, M., 'Recherches sur les armées hellénistiques', *Bibliothèque Des Ecoles Françaises d'Athènes et de Rome*, 196 (1951), 97–106.
Lee, J.W.I., *A Greek Army on the March* (Cambridge: Cambridge University Press, 2007).
Liddell, H.G., and Scott, R., *Greek-English Lexicon*, 9th edn, ed. H.S. Jones (Oxford: Oxford Univerity Press, 1961).
Loades, M., *The Composite Bow* (Oxford: Osprey Publishing, 2013).
Loades, M., *The Longbow* (Oxford: Osprey Publishing, 2013).
Lord, C., *Aristotle's Politics* (Chicago: University of Chicago Press, 2013).
Lynnerup, N., 'Cranial thickness in relation to age, sex and general body build in a Danish forensic sample', *Forensic Science International*, 117 (1–2) (2001), 45–51.
Mair, A.W., *Kallimachos: Epigrams* (1921).
Markle, M.M., 'The Macedonian Sarissa, Spear, and Related Armor', *American Journal of Archaeology*, 81 (3) (1977), 323–339.
Matthew, C., *An Invincible Beast: Understanding the Pike-Phalanx at War* (Barnsley: Pen & Sword Military, 2016).
Matthew, C., *A Storm of Spears: Understanding the Greek Hoplite at War* (Barnsley: Pen & Sword Military, 2012).
Matthew, C., *The Tactics of Aelian* (Barnsley: Pen & Sword Military, 2013).
McEwen, E., 'Persian archery texts: chapter eleven of Fakhr-i Mudabbir's Adab al-harb', *Islamic Quaterly*, 18 (1974), 76–99.
McGeer, E., *Sowing the Dragon's Teeth* (Wasington: Dumbar Oaks, 1995).
Mcleod, W., 'The Range of the Ancient Bow', *Phoenix*, 19 (1) (1965), 1–14.
Mellink, M., 'Archaeology in Asia Minor', *American Journal of Archaeology*, 78 (2) (1974), 105–130.
Miller, W., *Xenophon: Cyropeidia* (Cambridge: Loeb Classical Library, 1914).
Milner, N., *Vegetius: Epitome of Military Science* (Liverpool: Liverpool University Press, 1993).
Morgan, M., *Xenophon: The Art of Horsemanship* (New York: Dover Publications, 2008).
Murray, A.T., *Homer: Iliad* (Cambridge, MA: Harvard University Press, Loeb Classical Library, 1924).
Murray, A.T., *Homer's The Odyssey* (London: Harvard University Press, 1919).
Murray, S.R., Sands, W.A., and O'Roark, D.A., 'Recreating the Ancient Greek Javelin Throw: How Far was the Javelin Thrown?', *Nikephoros*, 25 (2012), 143–154.
Niese, D.A., *Peltasts and Javelineers in Classical Greek Warfare. Roles, Tactics, and Fighting Methods* (Dayton: Wright State University, 2012).
Norlin, G., *Isocrates: On the Peace* (London: Harvard University Press, 1980).
Oakshott, E., *European Weapons and Armour* (Martlesham: Boydell Press, 1980).
Pain, S., 'Arrow points to foul play in ancient iceman's death', *New Scientist* (July 2001).
Paton, W.R., *The Complete Histories of Polybius* (Cambridge, MA: Harvard University Press, Loeb Classical Library, 2010).

Perrin, B., *Plutarch's Lives* (Cambridge, MA: Harvard University Press, Loeb Classical Library, 1968).

Pignolet, F., Guillion, F., Duval-Beaupere, G., Guezard, B., and Tarriere, C., 'In vivo measurement of human weight supported by the successive anatomical level from C4 to the femoral head', *SAE (Society of Automotive Engineers) Technical Paper 902306* (1990), 13–21.

Plant, I. M., *Women Writers of Ancient Greece and Rome: an Anthology* (Oklahoma: University of Oklahoma Press, 2004).

Pritchard, D.M., 'The Archers of Classical Athens', *Greece and Rome*, 65 (1) (2018), 86–102.

Pritchard, D.M., 'The Armed Forces', in *The Cambridge Companion to Ancient Athens*, pp.405–421, (Cambridge: Cambridge University Press, 2021).

Pritchett, K., *Greek State at War: Part IV* (Berkeley: University of California Press, 1985).

Pseudo-Aristotle, *On Marvellous Things Heard.*

Quesada Sanz, F., 'Gladius hispaniensis: an archaeological view from Iberia', *Journal of Roman Military Equipment Studies, JRMES*, 8 (1997), 251–270.

Radice, B., *Pliny the Elder: Natural History* (Cambridge, MA: Harvard University Press, Loeb Classical Library, 1969).

Rankin, D., *The Celts and the Classical World* (London: Routledge, 1993).

Rhodes, P.J., *Aristotle: The Athenian Constitution* (New York: Penguin Books, 2002).

Rieu, E., *Homer: Iliad* (London: Penguin Books, 2014).

Rihill, T., *The Catapult: A History* (Yardley: Westholme Publishing, 2007).

Rihll, T., *Use of Classical Greek Light Infantry and an Investigation into the Cause of the Lack of* (2014).

Robinson, D.M., *Excavations at Olynthus. Part X. Metal and Minor Miscellaneous Finds. An Original Contribution to Greek Life* (Baltimore: Johns Hopkins University Press, 1941).

Rolfe, J.C., *Ammianus Marcellinus: Complete Works* (East Sussex: Delphi Classics, 1940).

Roller, D., *The Georaphy of Strabo* (Cambridge: Cambridge University Press, 2014).

Rose, C.B., *The Archaeology of Greek and Roman Troy* (Cambridge: Cambridge University Press, 2014).

Sabin, P., *The Cambridge History of Greek and Roman Warfare, Volume I: Greece, the Hellenistic World and the Rise of Rome*, ed. P. Sabin, H. Van Wees and M. Whitby (Cambridge: Cambridge University Press, 2007).

Sage, E.T., *Livy: History of Rome* (Cambridge, MA: Harvard University Press, Loeb Classical Library, 1935).

Schlake, D., 'Remember Alberta's Most Primal Bear Hunter?', retrieved 20 March 2019 from Wide Open Spaces website.

Sekunda, N., *Macedonian Armies after Alexander 323–168 BC* (Oxford: Osprey Publishing, 2012).

Sekunda, N., 'Macedonian Army', in *A Companion To Ancient Macedonia*, pp.446–471 (Chichester: BiblioLife, 2010).

Sekunda, N., 'Military Forces: Land Forces', in P. Sabin, H. Van Wees and M. Whitby (eds), *The CambridgeHistory of Greek and Roman Warfare, Volume I: Greece, the Hellenistic World and the Rise of Rome*, pp.325–356. (Cambridge: Cambridge University Press, 2007).

Sekunda, N., *Seleucid and Ptolemaic Reformed Armies 168–145: The Ptolemaic Army* (Vol. 2) (Reddish: Montvert, 1994).

Sekunda, N., 'The Sarissa', *Acta Universitatis: Folia Archaeologica*, 23 (2001), 13–41.

Silva, L., *Viriathus and the Lusitanian Resistance to Rome 155–139 BC* (Barnsley: Pen & Sword Military, 2013).

Skov, E.T., *Experimentation in Sling Weaponry: Effectiveness of and Archaeological Implications for a World-Wide Primitive Technology*, thesis (Lincoln, NE: University of Nebraska, 2012).

Slining.org.
Snodgrass, A.M., *Arms and Armour of the Greeks* (Baltimore: Johns Hopkins University Press, 1999).
Spence, I., Kelly, D., Londey, P., and Phang, S., *Conflict in Ancient Greece and Rome* (Denver: ABC-CLIO, 2016).
Spencer, W.G., *Cornelius Celsus: De Medicia* (Cambridge, MA: Harvard University Press, Loeb Classical Library, 1938).
Sperrazza, J., and Kokinakis, W., 'Ballistic Limits of Tissue and Clothing', *Annals of the New York Academy of Sciences*, 152 (1) (1968).
Spyridakis, S.V., 'Cretans and Neocretans', *The Classical Journal*, 72 (4) (1992), 4299–4307.
Sturdivan, L.M., Viano, D.C., and Champion, H.R., 'Analysis of Injury Criteria to Assess Chest and Abdominal Injury Risks in Blunt and Ballistic Impacts', *Journal of Trauma – Injury, Infection and Critical Care*, 56 (3) (2004), 651–663.
Sutherland, C., 'Archery in the Homeric Epics', *Classics Ireland*, 8 (May 2001), 111–120.
Szabó, G., Kiss, A., and Tóth, Z.H., 'A Shooting Experiment with Reconstructed Scythian Arrows and Bows : New Data on the Iron Age Siege at Dédestapolcsány-Verebce-tető', *Hungarian Archaeology*, 10 (3) (2021), 43–52.
Tadlock, S.K., *Poor in Life, Naked in Battle: Athenian Thetes as Psiloi in the Classical Age* (Raleigh: North Carolina State University, 2012).
Taplin, O., *Sophocles: Four Tragedies* (Oxford: Oxford University Press, 2015).
Taylor, M.J., *Antiochus the Great* (Barnsley: Pen & Sword Military, 2013).
Taylor, M.J., 'Sacred Plunder and the Seleucid Near East', *Greece & Rome*, 61 (02) (2014), 222–241.
Thomas, M.A., *The Role of Archers, Slingers and other Light-Armed Infantry in Greek Warfare from the Mycenaean Period to 362 BC* (Glasgow: University of Glasgow, 1987).
Toolbox, T.E., 'Density of Wood Species', retrieved 10 November 2017 from Engineering Toolbox website: https://www.engineeringtoolbox.com/wood-density-d_40.html.
Treister, M.Y., *Hammering Techniques in Greek and Roman Jewellery and Toreutics* (Leiden: Brill Academic Publishing, 2001).
Ulman, M.S., and Stalnaker, R.L., 'Evaluation of the AIS as a Measure of Probability of Death', *IRCOBI Conference*, 105–119 (Ohio: Ohio State University, 1986).
Walbank, F.W., *A Historical Commentary on Polybius* (Oxford: Clarendon Press, 1957).
Walbank, F.W., *The Cambridge Ancient History: The Hellenistic World*, 2nd edn, eds F. Walbank, A. Astin, M. Frederiksen and R. Ogilvie (Cambridge: Cambridge University Press, 1984).
Wardel, D., *Cicero on Divination* (New York: Oxford Univerity Press, 2007).
Warmington, E.H., *Remains of Old Latin* (Cambridge, MA: Harvard University Press, Loeb Classical Library, 1938).
Way, A.G., *Caesar: Alexandrian War. African War. Spanish War* (Cambridge, MA: Harvard University Press, Loeb Classical Library, 1955).
Webber, C., *The Thracians 700 BC–AD 46* (Oxford: Osprey Publishing, 2001).
Wells, B., *Diodorus Siculus, Library of History* (Cambridge, MA: Harvard University Press, Loeb Classical Library, 1963).
Whiston, W., *The Works of Josephus: Complete and Unabridged* (Peabody, MA: Hendrickson Publishers, 1987), 926.
White, H., *Appian's Roman History* (London: Loeb Classical Library, 1962).
White, J., *Bows and Spears in Achaemenid Persia* (Berkeley: University of California, 2019).
Wilkins, A.S., *M. Tulli Ciceronis Rhetorica* (Oxford: Oxford Univerity Press, 1903).
Wilkinson, R., 'The Representation of the Bow in the Art of Egypt and the Ancient Near East', *Journal of Ancient Near Eastern Society*, 20 (1991), 83–99.

Williams, M.F., 'Philopoemen's special forces: Peltasts and a new kind of Greek light-armed warfare (Livy 35.27)', *Historia - Zeitschrift Fur Alte Geschichte*, 53 (3) (2004), 257–277.

Wong, E., 'Abbreviated Injury Scale', in S.E. Bullard, M. Griss, S. Greene and A. Gekker (eds), *Encyclopedia of Clinical Neuropsychology*, pp.5–6 (New York: Springer Publishing, 2011).

Woodman, A.J., *Sallust: Catiline's War, The Jurgurthine War, Histories* (London: Penguin Books, 2007).

Yaden, Y., *The Art of Warfare in Biblical Lands* (New York: McGraw Hill, 1963).

York, R., and York, G., *Slings and Slingstones: Forgotten Weapons of Oceania and the Americas* (Kent, OH: Kent State University Press, 2011).

Zutterman, C., 'The bow in the ancient near east, a re-evaluation of archery from the late 2nd millennium to the end of the achaemenid empire', *Iranica Antiqua*, 38 (2003), 119–165.

Index